E. A. (Edward A.) Marsh

The Evolution of Automatic Machinery as Applied to the Manufacture of Watches at Waltham, Mass. By the American Waltham Watch Company

E. A. (Edward A.) Marsh

The Evolution of Automatic Machinery as Applied to the Manufacture of Watches at Waltham, Mass. By the American Waltham Watch Company

ISBN/EAN: 9783741166259

Manufactured in Europe, USA, Canada, Australia, Japa

Cover: Foto ©Andreas Hilbeck / pixelio.de

Manufactured and distributed by brebook publishing software (www.brebook.com)

E. A. (Edward A.) Marsh

The Evolution of Automatic Machinery as Applied to the Manufacture of Watches at Waltham, Mass. By the American Waltham Watch Company

THE EVOLUTION

OF

UTOMATIC MACHINERY

AS APPLIED TO

THE MANUFACTURE OF WATCHES

AT WALTHAM, MASS., BY

THE AMERICAN WALTHAM WATCH COMPANY

BY E. A. MARSH

———

WITH HALF-TONE ILLUSTRATIONS.

CHICAGO:

GEO. K. HAZLITT & CO., PUBLISHERS.

1896.

PREFACE.

An apology, to possess its highest value to all parties concerned, ought to be both prompt and unsolicited. The writer of the following pages desires, therefore, to apologize in advance for the short-comings and imperfections which may be found in this brief review of some of the steps of mechanical progress in the manufacture of watches on the American System. The work of preparing this brief history was performed in connection with the every-day factory duties of the writer and, therefore, subject to frequent interruptions and delays. It was not expected that it would be embodied in any more permanent form than in the columns of the monthly trade journal for which it was written.

E. A. MARSH.

Waltham, Mass., February, 1896.

INTRODUCTION.

As mankind develop in intelligence and culture, their wants become more numerous and varied and their requirements more exacting. The supply of one want but briefly anticipates the creation of a new one, which in turn demands satisfaction, so that the great and growing business of the world seems to be to supply its various wants. To a certain extent this supply business possesses the nature of a barter, in that it is an exchange of commodities, not always a direct exchange, indeed, it seldom is so simple a matter as that, yet in an indirect way all business as such is simply a channel through which the multitudinous wants of mankind find their supply. As the great majority of people have certain wants in common, the matter of providing an adequate supply becomes very important, and calls for special means or agencies through which to work, so that it follows that the better the means of supply the cheaper can be the supply obtained and the want satisfied. So that it is by no means an indication of laziness or indolence in an individual if he uses his brains in devising an easier, or quicker, or cheaper, or better method of performing his work. On the contrary, it will in many or in most cases prove quite the contrary fact.

It is one of the inherent conditions of human nature, especially of physical nature, that it is susceptible to fatigue. It is also a fact that a condition of weariness is not conducive to the attainment of the highest results, either in quantity or quality of work produced. If one of the qualities demanded in any certain kind of work be the highest attainable degree of uniformity, it will be readily admitted that the individual workmen, with the certainty of constantly recurring periods of fatigue, which make imperative corresponding periods of rest, is at a great disadvantage when in competition with an impersonal and tireless machine which is capable of producing work of a like kind. The man gets tired, or nervous, or is not feeling well, or is inattentive, or careless and indifferent. The machine has no such weaknesses, and though its work is not held up to the standard quality by any domination of its own conscience, yet its mechanical functions are so invariably exercised that its product of work will surpass that of its human competitor, not alone in quantity, but in exact uniformity as well. How much better then is a man than a machine? Within certain limits the machine is the

better. for the reasons already stated. Still, the powers
or capabilities of the machine are limited to a very nar-
row compass. i. e.. it can perform the particular duty
for which it was made, whereas the man. the human
machine, is capable of adaptation to the performance of
of a great variety of duties. But *within the field of its
capacity the properly designed and well made machine will
outstrip the man.* Especially will this fact be apparent
in the production of articles which from their nature
demand the exercise of special skill and unremitting care
and attention in their fabrication. Work of such nature
calls for a great degree of nervous strain on the individual engaged in producing it. The expenditure of nervous energy entails fatigue. which in turn reduces the
ability to exercise the needful skill or to concentrate the
operator's attention. and a lessened quantity and inferior
quality of product results.

If it is desired to produce large numbers of articles
which shall be substantially alike. and which are com-
posed of a variety of parts. it is evident that economy in
production will be best secured by a subdivision of the
labor. by which large numbers of similar parts shall be
produced by the same workman. It is also evident that
if the large number of required pieces whose function is
the same. can be made with dimensions exactly uniform,

there will result a great reduction in cost of manufacture because of the avoidance of any individual or special fitting of the various parts.

Such a system of manufacture has within the last fifty years come into very general adoption in all extensive concerns, and from what has been said, it will be evident that its success must depend upon the adoption of machines. and the comparatively small amount of individual skill demanded.   So far as is known to the writer. the most complete and comprehensive system of manufacturing on the interchangeable system. in its earlier years. was introduced by the United States Government for the manufacture of army muskets, at the United States Armory at Springfield, Mass.   Credit for a large share of the mechanical excellence of that system. and for the invention of many of the ingenious machines then and still in use for that work. belongs to the then master mechanic at the armory. Cyrus Buckland.   One of the interested students and great admirers of that ingenious mechanism in those early days was a young Boston watchmaker named Dennison. who was convinced that by the construction of suitable machines it would be not only possible but entirely practicable to manufacture the delicate mechanism of the pocket watch on the interchangeable system: and feeling that such a scheme was

feasible and might be made to yield a good manufacturing profit, he endeavored to establish a watch factory. to
be conducted on that system of manufacturing. After
years of delay he succeeded in organizing a company
and in beginning the work of preparation. As his plan
was to substitute machinery for skilled labor to a large
degree, it was of course needful to design and build the
special machines required. The field was new; experience could afford little assistance; money was by no
means plentiful, and it was therefore needful to proceed
with caution. so that expensive mistakes might be
avoided. It was. moreover. an uncertain matter as to
the magnitude of the business to be done. assuming that
the new manufacturing system should prove successful.
It is no detraction from the ability or the ingenuity of
the original mechanics of the watch factory that they did
not at that early date attempt the use of elaborate and
complicated machines, such as we now have in use.
Such an attempt at that time might indeed have given
evidence of inventive or mechanical ability, but it would
also have been an indication of poor business judgment.
and would have been a waste of the little money then
available. Large capital is an absolute necessity in the
manufacture and employment of automatic watch
machinery to any general extent. and in this branch of

business, as well as in almost all others, the larger concerns possess this great advantage over their smaller competitors.

If it were possible, it would be interesting to review the various forms of machines which have successively been used in watch making on the American system; which system, as has been suggested, consists in making large numbers of pieces of exactly uniform dimensions, so that they may be used interchangeably. As the only practical means of such manufacture, the system proposed and demanded special machines for the performance of the work, not only for the sake of economy, but also to do away with necessity for special skill in the individual workman. Such a review is, however, impossible. Most of the discarded or displaced machines have been destroyed. But of those which remain we may make a brief study, sufficient we trust to note the direction and path of progress in the evolution of the more complicated and costly machines now in use.

EVOLUTION OF AUTOMATIC MACHINERY.

CHAPTER I

In the review of machinery employed in watch making, it will be both natural and proper to consider. in the first place. the machines employed in the production of the plates. Of course the great bulk of the work on the plates is in the nature of turning, which involves the use of special chucks. and to a certain extent, of special lathes also. By the death of the late Ambrose Webster. the writer is deprived of an authority in certain matters of historical nature. for Mr. Webster was, during some of the early years of the Waltham factory, immediately engaged in the making of tools and machinery, and was well informed as to the origin of some of the foundation devices which have been universally adopted in American watch factories, and which by their evident value have also commended themselves to the attention of manufacturers in other lines. One of those primary devices, and perhaps the most important. is that of the " draw-in chuck." so called from the method of its

13

operation. it being caused to grasp the enclosed object
by being drawn into the tapering mouth of the lathe
spindle. It is also known as the "split chuck," from
the form of its construction. Probably the oldest form
of such chuck is found in the " Aiken's awl." (Fig. 1)
which is still in use in the
hollow awl handle, filled
with an assortment of awls
and other simple tools. This
awl handle is constructed
with a sleeve or socket. the
mouth of which is bevelled
or tapered, and the inner
end tapped with a screw
thread. The chuck, at its
inner end. is threaded. while
its outer end is bevelled to

Fig. 1.

correspond with the socket which receives it. The
projecting part is made square. to fit an accompanying
wrench. and the chuck being split nearly to the inner
end. forms two jaws which grasp the tool as the chuck
is screwed in. It is not now known whether in originally adapting this chuck to watchmaking lathes. the
projecting squared portion was retained. though it is
believed that it was so retained for a time, but was

succeeded by a solid rod or screw which extended through the axis of the lathe spindle and was operated by some sort of wrench applied to its outer end. This form of construction naturally suggested the use of a permanent handle. i. e., the "hand wheel" which is now in universal use. Still later the solid rod was succeeded by a hollow one, and the form of the closing chuck modified. until there was evolved the spring chuck and draw-in spindle. as now used in all watch-maker's lathes. Credit for the latter stage of improvement doubtless belongs to Mr. C. S. Moseley. who introduced it while the original of the Waltham watch factory was located in Roxbury, Mass.

Without doubt the early form of this split (or spring) chuck was used for holding wire, or small pieces of cylindrical form. but later the outer end of the chuck was enlarged. so as to hold the round discs of brass used for plates and wheels. But for holding work to be turned on its face, and which it was desired to duplicate or multiply in large numbers. and of exact and uniform thickness, the "draw-in chuck" possessed an inherent imperfection. If, in operating upon a succession of pieces, which it was desired to have of uniform thickness, one piece should be a little larger than another, the chuck could not be drawn into the taper mouth of

the spindle to the extent which the smaller pieces
would allow, consequently the larger pieces would be
turned thinner than the smaller ones. A similar effect
would be produced by unduly straining the draw-in
spindle. To overcome this inherent fault, some one
(probably Moseley), very ingeniously modified the
construction of the lathe so that the *chuck should be held
stationary,* as to any lengthwise motion, and the closing of the chuck be effected by the endwise movement
of the outside spindle. This form of construction,
although varied in some particulars, has been in
almost universal use in all watch factories for about
thirty years or more, and will doubtless continue. This
form of lathe is known as the "three bearing," or more
properly, "the slide spindle lathe." (Fig. 2.)

In 1873, Mr. C. V. Woerd designed and patented a
modified form of "slide spindle lathe," which was
nominally a *two-bearing lathe,* the third bearing being
obtained by making a long telescope fit of the draw-in
spindle in the rear end of the running spindle. With
very nice fitting, such a lathe would run fairly well for a
time, but it never was a satisfactory form of construction, and was abandoned. In the view of this lathe, as
shown (Fig. 3), it will be observed that it is made to
close the chuck without the use of the "hand-wheel,"

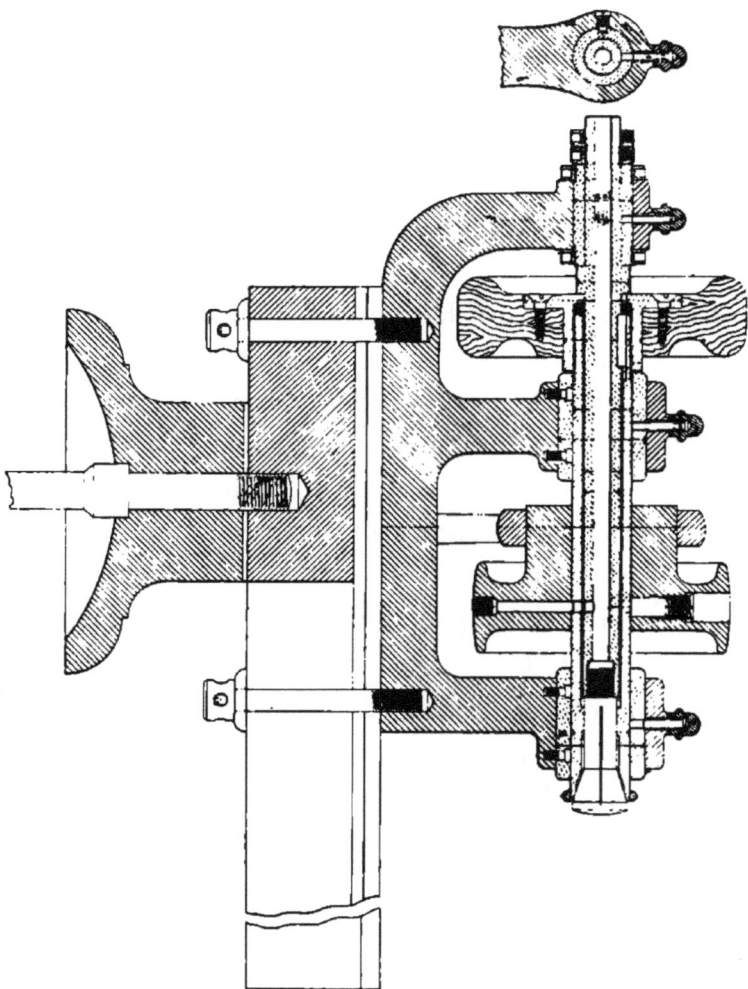

Fig. 2.—Three-Bearing, Slide Spindle Lathe.

the closing being performed by the action of a stiff
spiral spring. mounted on the draw-in spindle, while the
opening was done by means of a foot lever connected

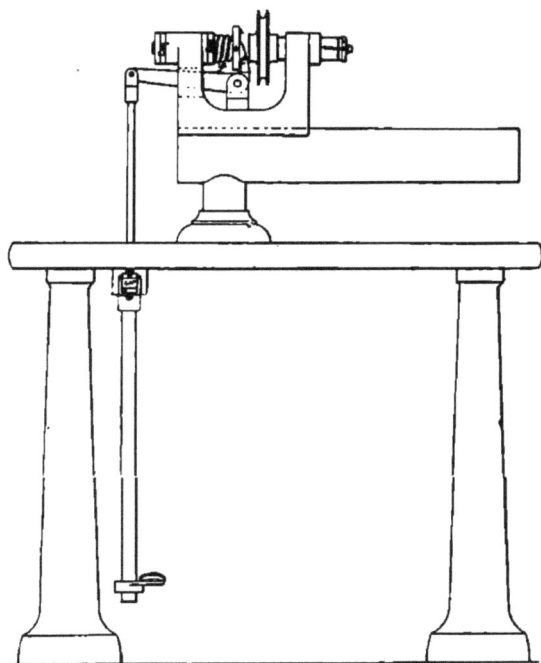

Fig. 3.—Woerd's Two-Bearing Slide Spindle Lathe,

with the forked bell crank lever attached to the lathe
head.   While this was not the original form of self-closing lathe. (as springs had for many years been used to
operate the draw-in spindle of small lathes for very light

work). yet it was an advance toward a better form of
self-closing lathe (Fig. 4). which has, by successive
steps. reached a form of construction so useful as to be
extensively employed. largely for the reason of its
adaptability to automatic operation, and. from that fact.
being peculiarly fitted to be incorporated in the more

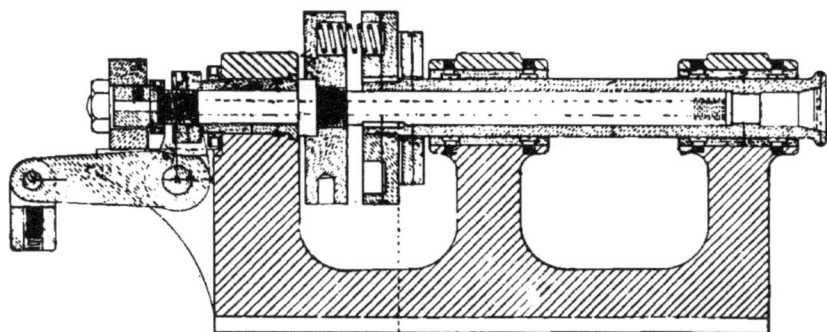

Fig. 4.—Self-Closing. Three-Bearing Slide Spindle Lathe.

complex forms of machinery which are made automatic
throughout.

Having thus briefly sketched the growth of the chuck
and the lathe. we reach the point of their joint operation
in the manufacture of watch plates.   It was the early
method in this factory to employ the slide spindle lathe.
above described. with hand wheel for closing and opening the chuck for holding the plates. in conjunction with
the common form of slide rest with the two slides, each

movable by screws turned by the two hands of the operator. Later the hand wheel was succeeded by the self-closing form of lathe. the opening of the chuck being performed by a combination of levers operated by foot. Then came the substitution of levers in place of the feed screws for manipulating the slide rest. Still later came an attachment to the lathe head for operating another tool. which served to round or bevel the edge of the plate after being faced by the slide rest tools.

A story is told of a young musical student who claimed that he could write a musical composition which should be correct. and yet be beyond the ability of his teacher to play. As his teacher expressed doubts as to the young student's ability in that direction, he was requested to submit his composition for trial. On laying it before his teacher for him to play on the pianoforte. there seemed to be nothing unusually difficult. until he came to a passage which demanded the use of the two hands at near the extremities of the keyboard. while there was call for a note to be struck on one of the keys near the center. Of course the professor stopped, with the exclamation that "such a passage could not possibly be played by any one." But the youth assured the master that it was not only possible

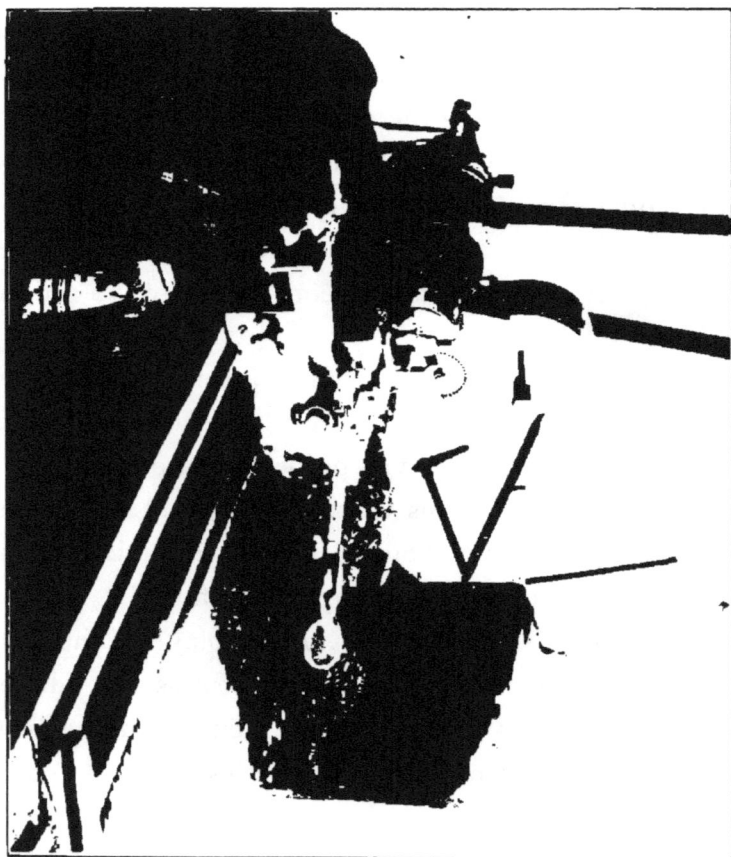

Fig. 5.—Plate Turning Lathe, with Double Lever Slide Rest and Cornering Tool.

of execution, but was by no means difficult, and seating himself at the instrument, he began the composition, and when he reached the peculiar passage he bowed his head and struck the required key with his *nose*. In the case of this plate-turning lathe, the operator's two hands are engaged in manipulating the slide rest, and his two feet in stopping and starting the lathe, and opening the chuck.   As a means of operating the cornering stool above mentioned, the *knee* of the operator is brought into service with perfect success.

It is obvious that a person could attend to but one lathe of the kind described, and that the watch plate would require a large succession of turnings, each of which would require a special chuck, consequently a very large number of lathes would be demanded, beside the consumption of much time in changing chucks and adjusting the tools.

Within two years, however, Mr. D. H. Church has constructed two machines for plate turning, which mark a new departure in that direction, the novelty being in their adaptation to the performance of a number of turnings, equivalent to a number of machines and operators. The first of these machines is designed for turning the recesses in pillar plates, such as are shown in the accompanying illustration (Fig. 6), which shows the train side

of a 16-size plate, with recesses turned for the barrel,
the escape wheel, the pallets, the balance, and the bearing for the intermediate setting wheel, also a small
recess for the center pinion.   The blank plates, having
been faced on both sides, and having the diameter
turned, and the dial feet holes made, are placed in a
tube at the left hand end of the machine, whence, they
are taken, one at a time,
by a swinging carrier arm,
and are placed in the chuck
of the first running spindle.
By the action of a cam on
a shaft below and parallel
with the running spindle,
which is at rest, one of the
slides of the compound tool
carriage is moved *toward*
the work, which carries the

*Fig 6.—Sixteen size Pillar Plate,
showing six recesses made by
Automatic Machine.*

turning tool into operative position, and the chuck spindle
being at the same time set in motion, the tool com-
mences its work of cutting the recess, beginning at its
outer edge.   When the tool has cut to the proper
depth, another cam comes into action, and the tool is
moved across the work, turning toward the center of
revolution.   If a slight boss is desired at the center of

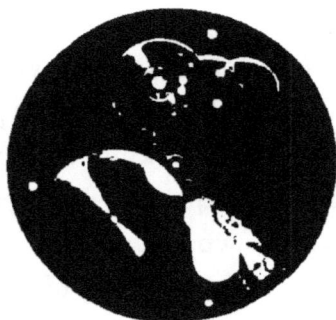

the recess. the tool is withdrawn at the proper time and
distance. and when the recess is complete the turning
tool is automatically drawn back and returned to its
original position, which allows room for a second carrier to swing over into position to receive the recessed
plate.   By this time the revolution of the chuck spindle
has ceased. and this second carrier moves up into contact with the plate. and as it recedes again it carries the
plate with it, and swinging over. it moves up and
deposits it in the chuck of the second spindle. but in
such position that the succeeding recess will be made in
its proper location on the watch plate.   Another and
suitable tool is now brought into action in a similar
manner to the first, and when its work is completed it
retires, and a third carrier takes the plate and places it
in another chuck.   In the meantime the two preceeding
chucks have received new plates. and so the work
progresses simultaneously, six recesess being made successively in as many plates. each unlike all the others in
size, position and form.

The last carrier in the series deposits the recessed
plate in a tube at the extreme right hand end of the
the machine,   By a change in chucks, etc., the turnings
on the dial side of the plate can be made in a similar
manner.   The boldness in the conception of this

Fig. 7.—Automatic Pillar Plate Recessing Machine.

machine. (Fig. 7) will be appreciated when it is realized that the watch plate must be placed in each succeeding chuck in a different position, and that it is required to be placed on three pins which fit in the three dial feet holes.

Another form of plate turning machine is designed for turning recesses in top plates, which are held in self-closing chucks, instead of being placed on pins as is the case with pillar plates. This machine, which is shown in the accompanying view, (Fig. 8) is exceedingly novel in its movements, especially in the action of the transfer carriers, which instead of swinging over, like those on the pillar plate machine, are made to swing around. This form of motion involves the necessity for an additional movement in a vertical direction to clear the tool carriage. In this case, as in the other, the blank plates are taken from a tube, where they have been placed by the attendant, and placed in the chuck, and are also delivered in another tube when completed. Another difference in action consists in the movement of the work toward the tool, instead of the tool toward the work. But the most novel and curious action is that of the transfer carrier. As this arm swings *around*, its ordinary effect would be to reverse the watch plate so that it would be placed in the succeeding chuck with

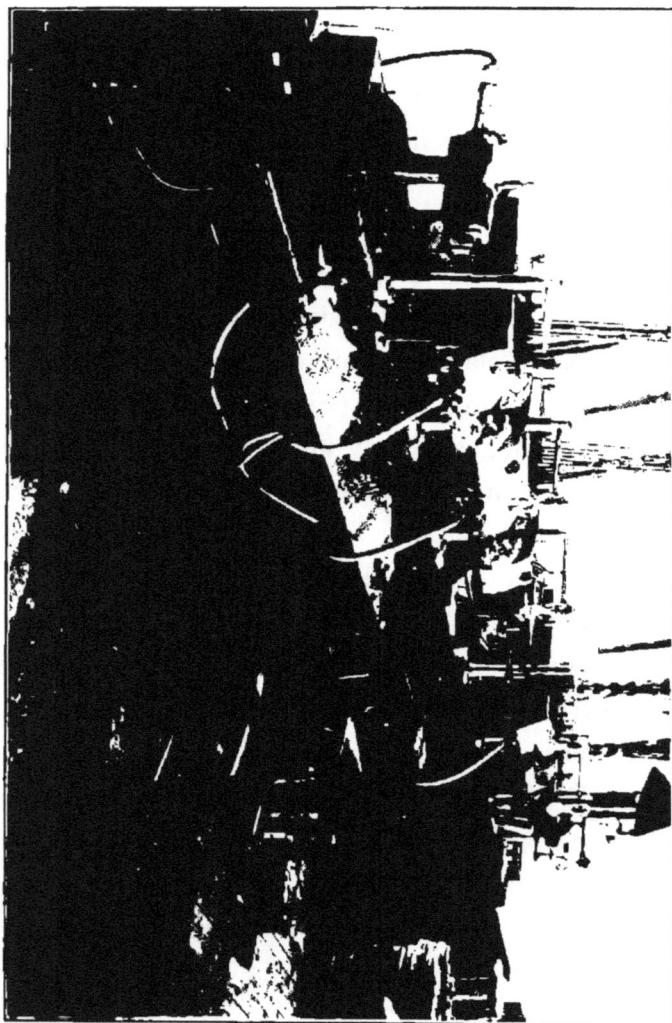

Fig. 8.—Automatic Plate Recessing and Facing Machine.

the opposite face presented to the cutting tool. In some
cases that change would be desired. but where it is
desired to make the successive recesses on the same
side. the carrier reverses the plate during its transfer,
so that it is placed in the second chuck with the same

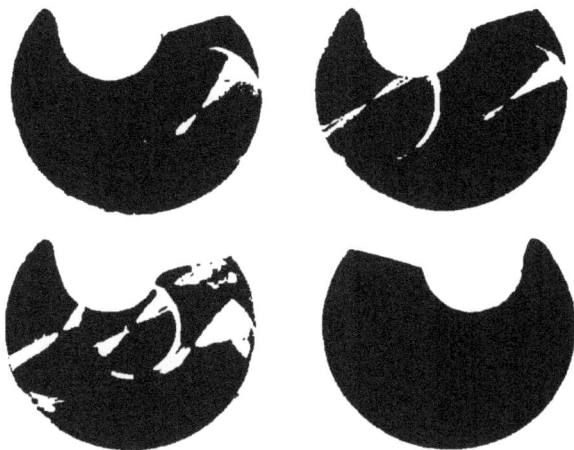

Fig. 9.—Showing Recessing and Facing of Top Plates.

side out. but with the center of revolution at what
ever point desired. The accompanying illustration,
(Fig. 9) shows the several recesses and the facing of
both sides of plates.

CHAPTER II.

It should be understood that this work is not intended to describe the manufacture of watches, for that would require volumes, but to briefly note some of the steps in the evolution of special machines, which characterize the "American system" of watchmaking. It must be understood that only a few of the more prominent machines can be considered, and that only a general description of those can be given.

Having briefly considerd the turning of the plates, we will defer, for a time, any mention of other operations on that part of the watch movement, and review the successive forms of the machines for turning the various arbors, staffs and pinions, which constitute some of the moving parts of the watch.

The history of the American Waltham Watch Factory does not extend back to that indefinite period when power was obtained by means of the foot wheel, and the turning tool was simply a graver held in the hand of the operator; but only to the time of the lathe and slide-rest, whose feed-screw was operated by hand. For a time, the turning of all staffs and arbors was performed in this

way.   But this method was succeeded by a form of
semi automatic turning lathe. the exact form of which
the writer is unable to learn.   Succeeding this, came
an improved form of lathe. in which the tool was caused
to move with a uniform speed and to a desired distance

Fig. 10.—Moseley-Webster Staff Turning Lathe.

and then automatically withdrawn from contact with
the work and returned to the point of starting.   The
accompanying illustration, (Fig. 10). shows one of the
oldest form of lathes used in this factory. which can be
found (and unfortunately minus one of its spindles).
It will be observed that at the left hand end of the

machine there is a combination of levers, designed to provide for turnings of various lengths without necessitating a change of feed cams.   It is the understanding of the writer that the combination levers were designed by Mr. Ambrose Webster, while the invention of the lathe should be credited to Mr. Chas. S. Moseley.

The next illustration (Fig. 11). shows a later form of automatic staff turning lathe. which embodies substanti-

Fig. 11.—Staff Turning Lathe (Vander Woerd Pattern.)

ally the same principle of mechanism as the earlier machine, but differently arranged. and of much heavier construction.   The cam shaft. at the back of the machine. is driven by a worm and worm-gear. not shown. At the extreme left hand end of this shaft is shown the feed cam. the acting face of which gives motion to a rack lever. which indirectly communicates motion (endwise) to the tool carrying spindle. one end of which is seen projecting through the end of the machine. The opposite end of this spindle enters the projecting cylinder at the right hand side of the machine. which cylinder contains a spiral spring, which serves to carry the spindle in the reverse direction to that imparted by the feed cam. About midway of the length of this spindle is attached the tool holder, which is compound in its construction, to allow for the desired elevation of the turning tool. and also to provide a means of moving the tool at a right angle to the axis of the staff or arbor to be turned, this latter motion serving to withdraw the tool from contact with the work, or. when so desired. to enable the tool to perform a " squaring out " or " facing " cut. To effect this motion an arm of the tool carriage extends back underneath the cam shaft and is acted upon by a suitable cam. which depresses the arm, and thereby slightly turns the feed or tool carrying spindle, thus moving the tool away from the work, and, while thus held back, the action of the feed cam allows the tool to return to its starting point, in readiness to act upon the next piece of work. Devices for so governing the action of the tool as to turn tapering staffs are also provided. but cannot be seen in a general view. In staff turning it is sometimes desired to have one of the spindles in which the work runs fixed and the opposite one movable. to allow of the reception and removal of the work, and at other times it is needful to make the opposite spindle.the fixed one. Ready provision is made for these requirements by means of suitable binders, or clamps provided with milled nuts, shown on the front of the spindle heads. When one of these nuts is screwed down so as to clamp the spindle firmly in place. the opposite one is turned back so as to leave the spindle on that side free to slide. When the staff to be turned is inserted in position. the free spindle is moved up into proper contact with it, and as the lathe is put in motion, a suitable cam on the rear shaft acts through an adjustable lever. and binds the slide spindle in place until the turning tool has completed its work and has returned to its starting point, when the spindle is released and can be slid back, and the completed arbor removed.

This brief, and perhaps not very clear explanation of the action of this staff turning lathe, is given for the benefit only of those readers who have had no opportunity of visiting a watch factory. As has already been said. this is substantially the form of staff turning lathe in use in American watch factories, and probably to some exent may have found its way into some European factory also.

Some years ago the writer designed another form of staff lathe, which, while in many respects similar to the

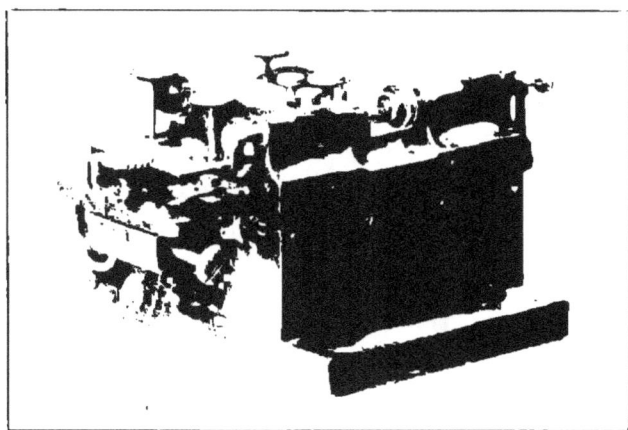

Fig. 12.—Marsh Staff Turning Lathe.

foregoing, yet possessed several features of novelty and special advantage. The accompanying illustration, (Fig. 12), will serve to give a general idea of its appearance. It was much more heavy and solid, and consequently assured a greater accuracy. It also had superior devices for taper turning in either direction, and in connection with such turning, it provided for facing either, hollow, crowning, or at a right angle to the axis.   It was also arranged to turn special pieces in which a part of the turning was to be straight and when a certain point was reached the turning continued at any desired angle.

Machines of all the above forms are to a large degree becoming obsolete in the American Waltham Watch factory, they having been largely displaced by the radical improvements designed by Mr. D. H. Church.   It should be said in connection with all the forms of staff turning lathes above described, that they are but semi-automatic.   Each piece to be turned, required to be " dogged " as a means of driving; the dog engaging with a suitable finger, or horn, on the driving whirl. The applying and removing of the " dogs " giving constant and lively employment to the attendant, and making impossible the running of more than a single lathe by one operator.

In the earlier days of American watchmaking the pinions were to a great extent made from " drawn pinion wire," but in later years the use of such special wire was discontinued, and the use of plain round steel wire was adopted.   For years it was customary to cut the plain wire into short pieces, making suitable allowance in length for finish.   The common way was to " chop off " the wire by means of an ordinary wire cutter or by a

special chopping machine. These short pieces were
then placed one by one in a spring chuck in a lathe, and
one end carefully pointed. After the entire lot had been
pointed on one end, they were placed in another chuck,
provided with a suitable interior stop, against which the
pointed ends of the blank should bear, in order to insure
exactness in the length of the blanks. Subsequent
" rough turning " of these blanks, which were still held
in spring chucks, removed a large portion of the surplus
metal, bringing them into a suitable condition to be
turned in the automatic staff lathes. When acted upon
in the staff lathes the blanks were held and revolved on
" dead centers," each piece as has been said, requiring
to be dogged by the operator.

Some of the smaller staff blanks, like pallet arbors,
were cut off in a lathe and one end pointed at the same
time, but all blanks, whether large or small, required to
be dogged at each individual turning. And, owing to
the minuteness, and consequent weakness of most of the
pieces, only a small amount of metal could be removed at
any single turning, consequently the required turnings on
some staffs were quite numerous, and in pieces such as
the balance staff, which contains numerous sizes and
shoulders, the number of individual turnings would be
ten or twelve, and in order to avoid injurious springing

of the staffs it was needful to alternate the turnings, so that the reduction in size should be gradual and uniform on both ends.

CHAPTER III.

In the preliminary, or rough turnings of the various staffs, a radical improvement was initiated by Mr. C. V. Woerd, about fifteen years ago, consisting of an automatic roughing out machine, which was so designed as to receive a rod of steel wire about twenty inches long, which was held by the ordinary self-closing chuck, whose evolution has been described. In suitable relation to the mouth or face of this chuck, was a turning and pointing tool, mounted on a movable carriage or frame which was actuated by a suitable cam, so that as' the wire rod, which projected the proper distance out of the chuck, revolved, this tool was gradually moved forward into contact with it, and cut away a portion of the metal and carefully pointed the end. When this operation was completed the cutting tool quickly retired, the revolution of the running spindle was arrested, the chuck loosened its grasp on the rod, which was then automatically fed forward a proper distance, and again grasped by the chuck, the spindle again started, and a second cutting tool moved forward, which immediately commenced its work of cutting off the steel rod; the severed piece to be of proper length to form a desired pinion or arbor. But just before the completion of the severing operation, a tubular carrier was moved over into axial line with the revolving rod and then moved back so as to enclose the severed piece, which, when entirely detached from the rod, was left in the carrier, which immediately moved forward and out of the way. The running spindle was again stopped and the chuck released its hold, and the rod was fed forward again, and the first tool was again brought into action, and so this part of the work proceeded. In the meantime, the above mentioned carrier was moved over into axial coincidence with the second spindle whose chuck was in readiness to receive the severed blank contained in the carrier. Immediately the carrier reached its position in front of this second chuck, a push rod moved forward and entered the rear end of the carrier tube and forced the enclosed blank into the mouth of the second chuck, which at once closed upon it. The carrier then retreated a short distance, the push rod was withdrawn and the carrier moved to an intermediate position to await its proper time to secure a second blank from the first chuck. As soon as the carrier was out of the way, the second spindle was revolved, and a cutting and pointing tool began its work of forming the second end of the blank, and when its

Fig. 13.—Woerd-Marsh Automatic Roughing Machine.

work was completed the spindle was stopped, the chuck opened and the completed blank ejected and fell into a chute. which deposited it in a receptacle entirely separate from the cuttings, so that the work of separating blanks from chips was entirely avoided.   This machine was so arranged that when the wire rod had been entirely converted into blanks, the machine would stop itself, and, unlike some human machines, it would not go through the motions unless it was actually doing work.   Inasmuch as these machines were so completely automatic in action, and the wire rods were of a length capable of being cut into a large number of pieces, a single attendant could care for the running of six or eight machines.   The establishment of these automatic roughing machines (Fig. 13) served to greatly reduce the cost of staff turning, but there still remained the numerous " finish turnings " to complete the pinions or staffs; and. as has been said, each individual turning required the application of a driving dog.   To supplement the work of this machine a radical improvement has been made by Mr. D. H. Church, who has invented and patented an automatic turning machine. which is really a complete battery of staff turning lathes which are located on a single bed or table. and all of whose operative mechanism is driven by a single belt.

A detailed description of any of these complicated machines would be somewhat difficult, and would be uninteresting to the general reader, and will not be attempted. It may be briefly said that at one end of a long bed or table is located a suitable frame, provided with a vertical shaft which at its upper end, carries a disk or plate. On the face of this plate, and near its edge, are turned suitable concentric grooves which are crossed by radial V shaped grooves of proper size and equally spaced. In these grooves are uniformally placed the roughed out blanks, from which are to be formed arbors or staffs. Adjoining this blank holding device is located an automatic staff turning machine, and at uniform distances beyond it are similar machines, sufficient in number to perform all the required turnings on any given staff. Alternating with these machines are upright stands or columns, through whose centers project spindles or shafts. From near the top of these upright shafts extend arms, from which, at their outer ends depend suitable clips or fingers. These shafts have reciprocating motions in both rotary and vertical directions. The foundation device, that which makes this machine entirely automatic, is the one for grasping the blanks while being turned, so that the applying of a separate " dog " is rendered unnecessary. A very

important gain is also obtained by this method of driving. viz.: the ability to remove a much larger amount
of metal at a single turning than could be done with
the ordinary method of dogging. so that a single
attendant is able to produce as many finished pieces as
would six. or more. people under the old system.   We
have mentioned. above. that this machine was arranged
to do the finish turnings on staffs or arbors which had
been roughed out on another machine. but by the substitution of a cutting off head for the above mentioned
blank holding plate. a class of work which requires the
removal of a smaller amount of metal can be turned
direct from the wire rod. which may be of considerable
length. say five feet.   The accompanying illustration
shows the machine arranged in the latter form. (Fig. 14.)

In operation, a long piece of suitable steel wire is
placed in the tube. extending to the left of the machine,
with the inner end of the wire projecting slightly from
its holding chuck.   The machine is then started, and
a suitable cutting tool advances and carefully turns the
projecting wire to a suitable point.   Then by suitable
mechanism the revolution of the spindle is stopped, the
tailstock spindle is moved toward the chuck to a definite
point, the chuck is opened and the pointed wire fed forward till it comes in contact with a suitable center in the

Fig. 14.—Mr. Church's Battery of Automatic Staff Lathes.

tailstock spindle. the chuck is then closed. the tailstock
spindle is clamped firmly in place. the spindle is revolved.
and the cutting tool moves forward and begins its work
of severing the wire.   Just before the piece is completely
severed. the arm of the upright shaft swings around so
as to bring the clip directly over the piece.   It then
moves down. and the fingers grasp the then severed
blank. the tailstock spindle recedes so as to clear the
blank. which is then lifted clear of all obstructions and
carried around exactly 180 degrees. and then again low-
ered to a position exactly between the centers of the next
machine. which centers then advance and close upon the
blank.   The fingers are then lifted out of the way. a
driving clamp closes upon the blank. the spindle is
started. and a suitable cutting tool moves up and com-
mences its work of turning.   When this turning operation is completed the tool is withdrawn, the spindle is
stopped. the driving clamp is loosened. and the partly
turned blank left clear of all obstruction. when a second
arm is swung over and another set of fingers descend
and grasp the blank, and, lifting it. gets it out of the way
just in season to allow a second blank to be inserted,
which is then treated in the same manner as the first one.
It will be understood that in carrying the blank from
one machine to another, the blank is reversed as to its

relation to the head and tail stocks and the cutting tools,
the second carrier placing the blank in the third machine
so that its opposite end is to be acted upon by the cutting tool. When the work in the third machine is completed, the blank is in the same manner passed along
to the fourth machine, and then the fifth, which completes the turnings, and the next carrier drops the completed blank in a proper receptacle. It must be understood that the completion of the turnings in four
machines is made possible by the fact that the machines
are so arranged that more than one turning is performed
in a single machine. It will also be understood that all
the machines are in operation simultaneously, each on
its own blank, so that after the first blank has reached
the receptacle the procession is constant until the rod of
wire is completely gone: so that a completed blank is
dropped into the dish at the last machine as often as
another one is severed from the wire rod at the first
one—time about fifteen seconds.

We have already explained that the operation of this
machine is made possible by the invention of the automatic dogging or driving device. This device was, however, first applied to ordinary or isolated staff turning
lathes, which were modified to adapt them to this
improvement. In the first form of such adapted lathes,

the blanks were placed in a suitable hopper. from which they were taken. one at a time. automatically. This form of machine is in quite extended use also.

In all the foregoing forms of turning lathes, the application of some form of dog or driving device is an indispensable feature; but within a year Mr. Church has per-

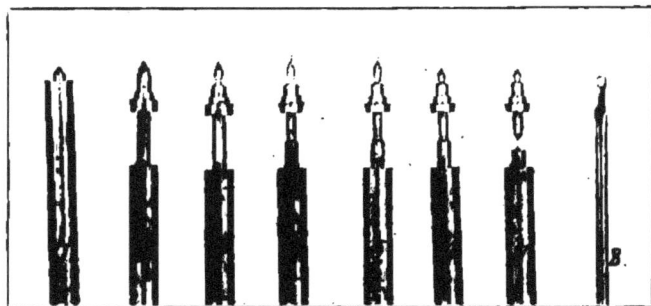

Fig. 15.—*Progressive Steps in Making a Balance Staff.*

fected a new form of machine in which that feature is rendered needless, and in demonstrating that fact he adopted the most difficult, delicate and complicated staff in the whole watch movement, viz., the balance staff. These are now made complete, every turning on the entire staff. including both pivots, being done at the rate per machine of 400 per day. or one staff each 90 seconds. We believe that nothing in the way of turning has heretofore been done which could at all compare with the work of these machines in delicacy, complexity and accuracy. (Fig. 15.)

The illustration will serve to indicate some of the progressive steps in the production of this complicated staff, and will also furnish an excellent specimen of microphotography, credit for which belongs to Mr. H. E. Duncan, who is known to many watchmakers.

To better indicate the extreme delicacy of this work, a No. 9 or No. 10 sewing needle is photographed in the same group, and serves to show the relative size of the two articles. Unlike the previously described machine, this machine operates only upon a single blank at a time, completing one blank before beginning upon another, whereas the former carried on all the successive turnings simultaneously, there being as many blanks in progress as there are separate heads to the machine. (Fig. 16.)

The accompanying illustration shows the appearance of two of these balance staff-making machines, but the large number and complication of the different movements required, render it difficult of description; we will therefore make no attempt in that direction.

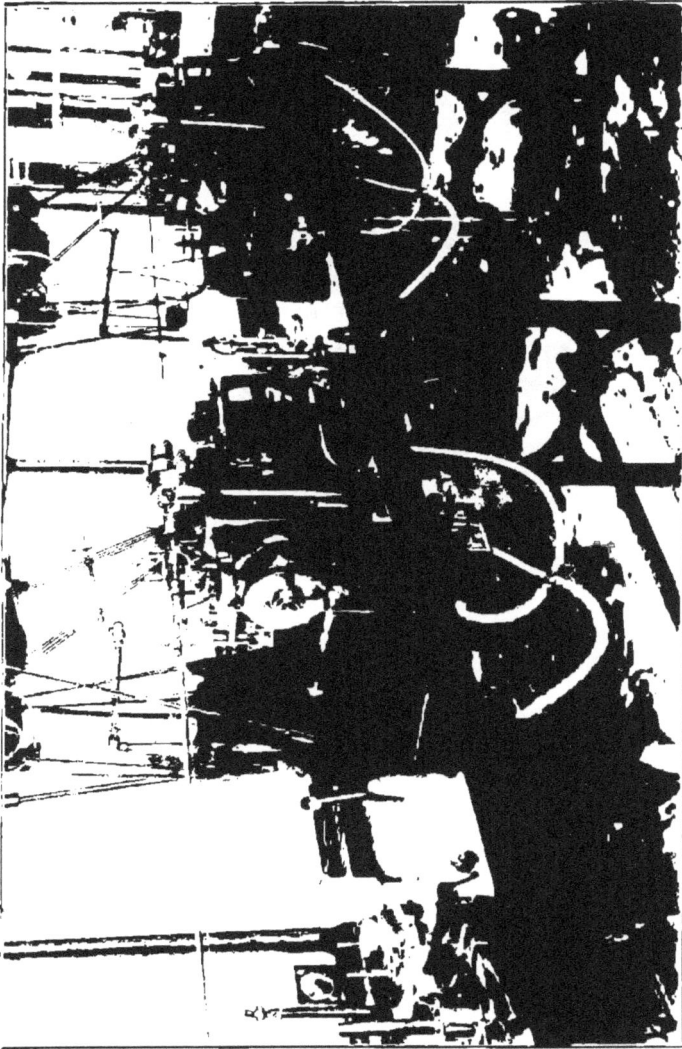

Fig. 10. Balance Staff-Making Machine.

CHARLES S. MOSELEY.

CHAPTER IV.

To a person who is familiar with the machines now in common use for the cutting of the teeth of wheels and pinions, the means for the performance of such work which were in early use would seem exceedingly crude and unsatisfactory. Possibly they may have been so regarded at that time. but it must be born in mind that the crude appliances which were first used did serve to produce material sufficient in quantity, and of enough excellence in quality to demonstrate that American watchmaking was possible. Improvement in quality and facility in manufacturing were but matters of time and money.

Mention has been made of the fact that pinions were made from specially drawn wire, in which the number and approximate form of the teeth, or "leaves" was given by the drawing dies. This wire was imported. and was received at the factory in pieces about twelve inches long. These were cut into lengths desired for the various pinions. the ends pointed. and the staffs turned. Then came the cutting of the leaves. The form of machine first employed for this work is unknown to the writer, but he recalls the fact that one of the first jobs of work which he did at the Waltham factory was to make drawings for the remodelling of two of these pinion cutters. But as

Fig. 17.—Old Style Pinion Cutters.

the work had already been commenced. he is ignorant of the original form of the machines. We have, however, preserved these machines in their improved forms, which are shown in the two accompanying views, and are so placed as to show the mechanism which was largely automatic in action (Fig. 17.)

It will be observed that this machine has no index for the spacing of the leaves, so that it is a matter of surprise that a safe degree of accuracy could have been attained, but we believe there was seldom any serious trouble from that source. Aside from the fact of these machines being very ingenuous (as will be evident from examination of the illustrations), they afford very good examples of the relative size of machines of those earlier dates, as compared with machines now constructed. These machines occupy a space of considerably less than six inches each way, while machines for similar use as now made, would occupy at least four times as much space, and be proportionately heavy. Just here it may be proper to say that examination of a variety of the earlier forms of the American watch machinery makes it very evident that the idea then obtained that delicate machines were necessary for the manufacture of the delicate mechanism of the watch. This idea was, after the experience of a few years, found to be a greatly mistaken one. But the writer recalls that very early in his connection with the Waltham factory, as he had made a drawing of some new machine, Mr. Ambrose Webster, then master mechanic, said to him, " You are running us to cast iron." But certainly there has been since that time no tendency to return to the practice of building light machinery. At the time of which we are writing the Waltham factory was run under two almost distinct

departments, the " Full Plate Department," under the
general charge of Mr. A. T. Bacon, with Mr. A. Webster, at the head of the mechanical department: while
the three-quarter plate movements were under the super-
vision of Mr. Chas. W. Fogg, with Mr. Chas. Vander
Woerd in charge of the mechanical work. This arrangement resulted in the establishment of a branch machine
shop in connection with the three-quarter plate department, and gave Mr. Woerd the opportunity to exercise
his inventive ability in designing and building some
special machines, of which mention will be made at
appropriate times. One of his earlier machines was
an automatic pinion cutter, which was quite ingenious,
and also somewhat complicated in action. It used to be
remarked that one who was not familiar with this
machine could get it into a " snarl " quicker than any
machine known. But, when understood, those machines
were capable of doing good work and in good quant-
ities. These machines were provided with three cutter
spindles, mounted in a revolving head, which were successively brought into action, so as to form the pinion
leaves from plain round wire instead of the English
pinion wire. We are unable to present an illustration of
those machines, for the reason they have all been somewhat remodelled within a few years so as to make

them entirely automatic in action. We will therefore defer further mention of them at this point. but will speak of them hereafter. Shortly after the introduction of the three-spindle machine above mentioned, Mr.

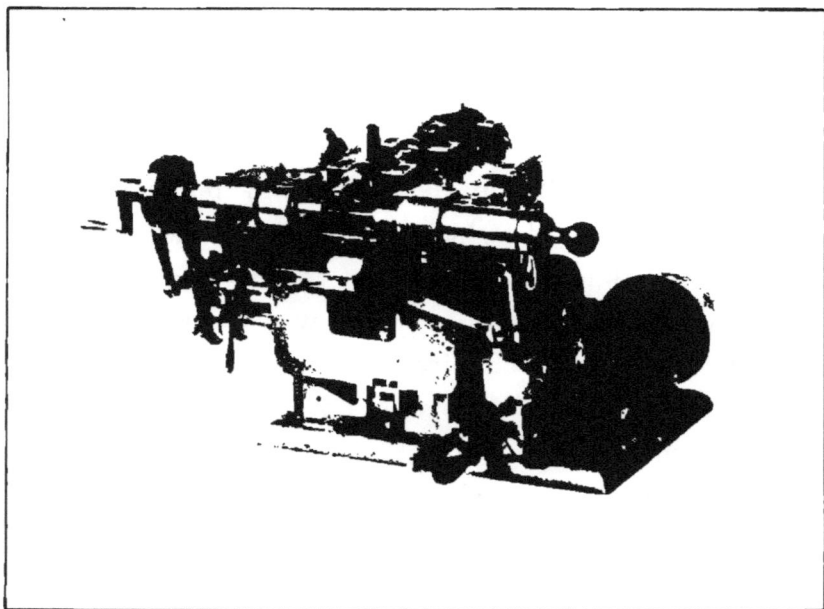

Fig. 18. —Improved Pinion Cutter.

Webster suggested the plan of making a pinion cutter in which the three cutters. instead of being on separate spindles. should be mounted on a single running spindle which should have an endwise movement so as to bring the several cutters successively into operative position. Such machines were made. and subsequent alterations and improvements brought them into the form shown in the next view (Fig. 18.)

These have proved to be very serviceable machines. and especially adapted to cutting a certain class of pinions which are more difficult to cut than others.   In the original Woerd machine there was no provision for adjusting the individual cutters to depth, other than could be made by variation in the relative diameters of the cutters themselves, which rendered the desired accuracy an attainment of a good deal of difficulty.   But the single spindle machine was provided with means for *adjusting each cutter* in two positions, so that although the cutters are fixed in their relation to each other on the running spindle. yet in operation they are entirely independent. The single spindle machine has proven to be most excellent in plan. and has been adopted in subsequent forms of pinion cutters, and has to a certain extent been copied by builders of machinery for use in other factories.

It is generally a fact that an individual who is to any considerable extent. engaged in the production of articles of any given nature. will naturally have some favorite form of construction.   Particularly is this true

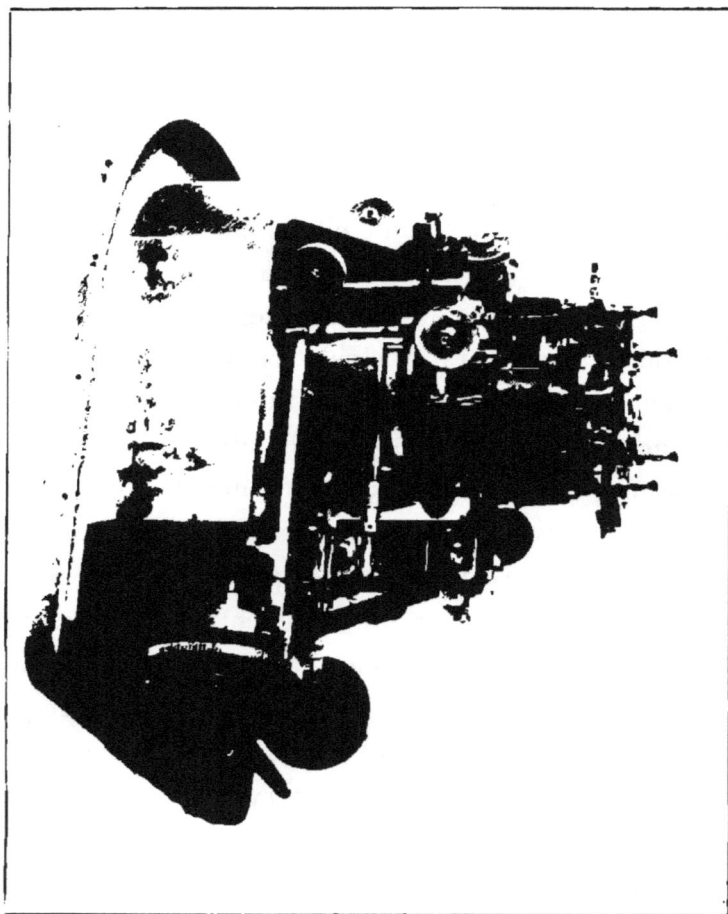

Fig. 19.—Revolving Cannon Pinion Cutter.

in the designing of machinery. It has for many years
been a theory of the writer that in the production of
large numbers of pieces, or articles of uniform dimensions, the economical method of manufacturing is to
"maintain a procession," and a procession to be end-
less must of necessity proceed in a circle. Circular
machines have therefore been a favorite form. The
most serious objection to such a form of construction
lies in the essential fact, that a great degree of accuracy
of workmanship is required. But if such accuracy can
be secured there can be no question of superior
efficiency.

The next illustration (Fig. 19) shows a machine of
this nature which was used for several years in cutting
cannon pinions. In this machine there was a plurality
of spindles for the holding of the pinions. Each of
these spindles was provided with an index, and, as they
were arranged around a common center, the entire
number were operated simultaneously. After the
indices had completed a revolution, the spindle-carry-
ing-head was caused to automatically make a partial
revolution, which brought each spindle into operative
relation to another tooth-forming-cutter. These cutters,
to the number of three or more, were mounted on
double slides, and were moved in unison, the cutters

performing their work, then being withdrawn and
returned to their former position, and then moved forward, and, the index having in the mean time moved
one space, the cutter advanced and cut a second tooth.
After all the teeth in the pinion had been cut, or blocked
out, by the first cutter, it was presented to the action of
the second cutter, as has been said, and then to the third,
and, if desired, to a fourth also. As this machine had
one more work spindle than it had cutters, there was
always a vacant spindle into which the operator could
insert a new blank. And as fast as *one cutter* could
perform its work on the pinion presented to it, just so
fast would a pinion be completed and removed, and its
place be filled with a fresh blank. The objectionable
feature in this form of machine has already been stated,
viz.: the difficulty of obtaining sufficient accuracy in
construction to secure the absolute "tracking" of the
several cutters. To overcome this difficulty, and at the
same time produce a machine of great productive
capacity, the machine shown in the next view was made.
(Fig. 20). In this machine another advanced step was
taken, by the incorporation of the self-feeding feature.
This machine is practically the combining of eight
machines in one, and is capable of cutting pinions
having 7 to 12 leaves; indeed, several kinds may be

cut at the same time. A brief description will serve
to explain the operation of the machine. which as is
shown in (Fig. 20) is circular in form. and consists of a
central column supporting a circular bed. on which is
placed a corresponding carriage. on the upper surface
of which are eight radial dovetailed grooves in which
move a like number of slides each of which carries a
complete head and tail stock, the tail stocks being
located on the inner end of the slides.   At one side of
each of these slides is mounted a frame carrying a
running cutter spindle.   These frames are capable of
a sidewise motion, so as to bring successively into
operative relation to the pinion blank each of the three
cutters on the cutter spindle.   Suitable stops are provided for the accurate adjustment of each of these cutters: both as to depth and position of cut   Directly
over each of the slides which carry the head and tail
stocks. is located a suitable magazine. or pinion blank
holder. open at the bottom, where are arranged suitable
elastic fingers designed to grasp a single pinion blank.

An upright shaft. located in the center of the machine.
gives motion to the entire mechanism.   The operation
of the machine may be described briefly. as follows:
The various cutters being properly adjusted, and the
magazines loaded with blanks. the shaft is started. and.

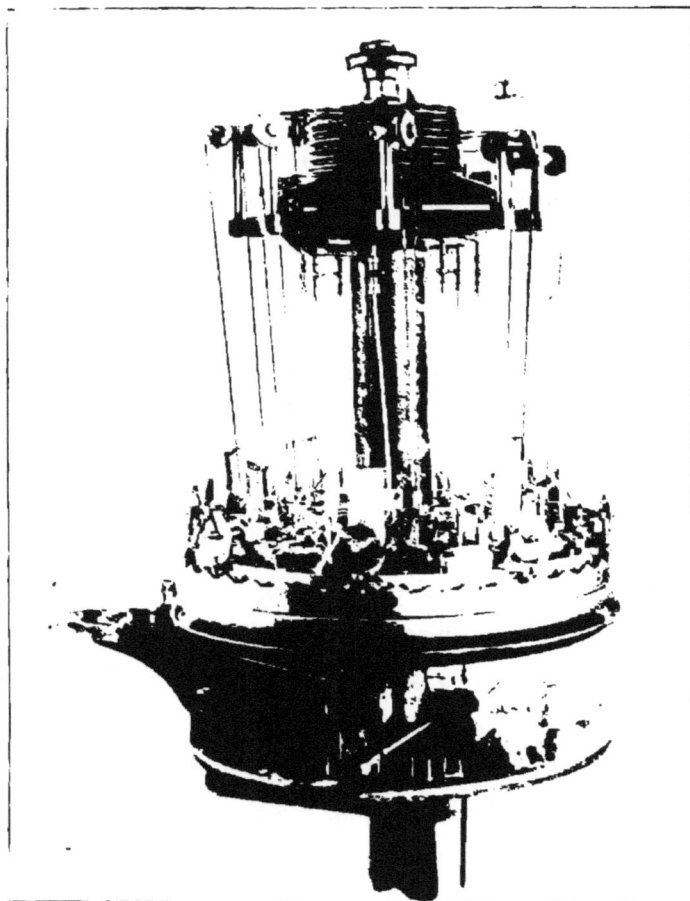

Fig. 20.—Marsh's Continuous Self-Feeding Pinion Cutter.

through the several belts running from the multi-grooved pulley at the top of the machine to the several cutter-spindles. they are put in motion. The circular table begins to travel. carrying all the radial slides above mentioned. When a certain point in its orbit is reached. one of the magazines descends until the axis of the pinion blank. which is grasped in the fingers at its lower end, comes in exact line with the centers of the head and tail stock of that slide. Both the head and tail stock spindles advance toward each other until they close upon the waiting blank, the taper staff of which is forced into a suitable socket in the head stock spindle. The magazine now rises, and as it loses its hold of the pinion blank, it receives into its grasp a second blank which it continues to hold until it shall arrive at the place of deposit, as before.

Immediately the magazine has taken itself out of the way, the slide begins to move in a radial direction in relation to the revolving. thus carrying the pinion blank back and forth over the revolving cutter. As the slide reaches its limit of outward motion. the index on the head spindle is revolved one division. and during the return motion a second tooth is cut. When all the teeth in the blank have been blocked out by the first cutter, the frame which carries the cutter spindle is moved sideways, so as to bring a second cutter into acting position, which proceeds with its work, and then gives place to a third and final cutter, which completes its work just a little before the revolving table has reached its starting point. The tail stock spindle is now withdrawn, and a little lever on the side of the stationary bed of the machine springs forward just as the outer end of the head spindle is passing it, and quickly drives the completed pinion from the socket, which is then ready to receive a second one. The magazine now descends and the second blank is taken, as before.

It will be understood that while we have been following this first blank in its travel and progress, each of the seven other slides has received its blank, and those are also in all stages of progress, from the first cut of the first tooth to the finishing cut of the tast tooth. The procession is uniform and continuous, and when each of the eight radial slides, in its continuous travel, reaches a certain point in its orbit, it delivers a completed pinion, and immediately receives another blank and continues on its way. Unlike the previously mentioned machines, these blanks do not pass from one part of the machine to another to be acted upon by local cutters, but the group of three cutters accompanies the blank on its travel. This arrangement permits the cutting of different kinds of pinions to be performed simultaneously. In this machine the work of the operator is reduced to the occasional examination of the completed pinions, and the supplying of the magazines with blanks. Simple devices are provided for rendering each of the slides inoperative whenever so desired, without in any way affecting the action of the others.

The success of the automatic feeding of the blanks and discharging of the finished pinions led to the adoption of equivalent devices on machines formerly in use. The next illustration (Fig. 21) shows a long row of *remodelled machines* of the Woerd type previously mentioned. As these machines are in a row, the attendant, who is able to care for six or seven machines, is obliged to pass from one to another. To permit such movement and, at the same time, to avoid the fatigue incident to long hours of standing, a track is laid on the floor in front of the machines, and chairs are provided with grooved rolls which follow the track, allowing the attendant to glide easily and quickly the entire length of the group of machines under her care. In contrast to this method the previously mentioned machines regularly *presented themselves* to the attendant whose seat was stationary. Since the inauguration of

the foregoing machines, another form of machine has been also put in use, which embodies some of the features already noted, but which substitutes a different and improved form of automatic blank feed. These machines, by the employment of a large amount of oil, are able of working more rapidly than any heretofore used, and are also adapted to the cutting of wheels of all kinds, and will be considered in a subsequent chapter.

CHARLES VANDER WOERD.

CHAPTER V.

In the last chapter we remarked upon the apparent crudity of many of the machines in early use in the original American watch factory. They were primitive. certainly; but the conditions under which they were made were of the same character, and the fact of such tools being made and used is not to be taken as evidence of a lack of inventive ability on the part of those who used them, but rather as showing an ability to accomplish desired results by the simple means to which a lack of capital restricted them.

The limited market for the early product of American watches, would of necessity forbid the expense of building special machines of restricted character or capabilities. and compel the use of appliances of limited cost. and of a nature adapting them to a variety of uses.

In this paper we will consider some of the numerous and successive forms of machines for the cutting of teeth in watch wheels of brass. The earliest of these; within the knowledge of the writer. was a machine for cutting train wheels. and was simply a small iron planer such as is used in machine shops, or by model makers.

It had the ordinary reciprocating bed. or table. which was moved back and forth by means of a hand crank, the connection probably being by the old-fashioned chain. such as was formerly used in the feeding of the

Fig. 22.—Webster Train Wheel Cutter, Improved.

carriages of engine lathes. On the cross head of this little planer was mounted a suitable frame which carried the fly cutter spindle, which was driven from above by belt in the ordinary manner. On the movable table was placed the head stock. whose index was operated by hand. While this arrangement was somewhat crude. it served its purpose. and had its day. But in 1865 Mr. Ambrose Webster designed a new and excellent machine for train wheel cutting. with mechanism for automatically moving the carriage. and also for operating the index, (Fig. 22) which, in an improved condition, is shown in the accompanying view. The most serious fault in this machine was the manner of adjusting cutters for depth. which adjustment was effected by means of an eccentric quill. in which the cutter spindle was journaled. It would probably be a true statement of fact to assert that this machine has cut more millions of watch wheels than any other machine in the world. it having been in almost constant use for thirty years. Following this. and possessing several advantages over it. was one designed by the writer about 1872. which is shown in the next view. (Fig. 23.) As this machine was, at a later date. modified so as to adapt it to the cutting of steel wheels. we will defer further mention of it at this time, and consider some of the earlier forms of machines for brass wheel cutting.

The next illustration shows a machine with a history; more of history than will ever be written. It is doubtful if there is a person living who knows the

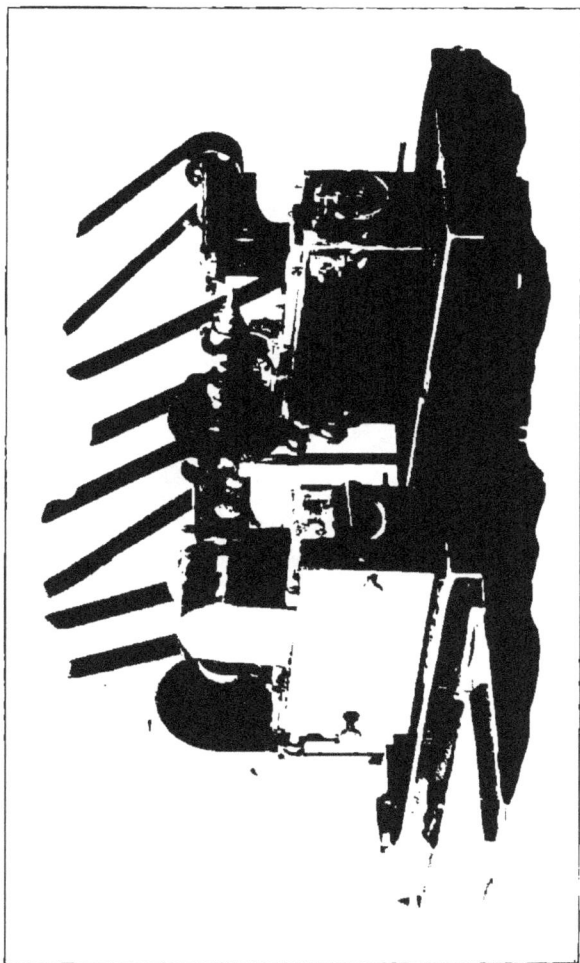

Fig. 23.—Marsh Train Wheel Cutter.

original form of this machine (Fig. 24.) or one who could accurately narrate the numerous changes, additions. and improvements which were made in it. Within the knowledge and recollection of the writer, this machine was used for cutting the teeth of minute

Fig. 24.—Old Minute Pinion and Hour Wheel Tooth Cutter.

pinions and hour wheels. and it would probably be safe to assume that it was originally designed for such use. The work was placed on an arbor at the top of the vertical quill spindle. shown in front, and held in place by a yoke which served as a sort of tail stock. The fly cutter was mounted in a short arbor or spindle. which revolved on centers. and also carried the belt

pulley. Reciprocal motion of the cutter was obtained
by the vibration of a frame. which was mounted in
suit·ble bearings on an adjustable carriage. which could
be moved toward or from the stationary quill. for the
purpose of adjustment to size. On this vibrating cutter-
frame was mounted an adjustable arm which extended
beyond its axis and terminated in a sort of toe. against
which pressed a cam. which was mounted on a driving
shaft: on one end of which shaft was a loose running
clutch pulley. while on the opposite end was another
cam. which. through a system of levers. slides. springs.
and catches. performed the operations of unlatching and
turning the index. It will be observed that the path of
the cutter through the work. instead of being in a
straight line, described a short arc of a circle. Of course
such a motion involved a theoretical imperfection in
work produced. but inasmuch as only a single thin
wheel or pinion was cut at a time. the error was indis-
tinguishable. This machine just failed of being automatic in action. for the mechanism always stopped before
the last tooth was cut. but as the driving of the fly cutter
was independent of the other movements. the operator
was able to depress the swinging arm for the cutting of
the last tooth. When in operation. this little machine
produced an impression that something was being done,

for the combination of noises caused by the operation of the various slides, clicks and latches, mingling with the hum of the fly cutter, was quite unlike any other combination in the factory. But it accomplished a good deal of work, and if it were human we would give it a

Fig. 25.—Improved Minute Pinion and Hour Wheel Tooth Cutter.

most respectful salute. The foregoing machine was some years since displaced by the machine shown in the next view (Fig. 25), which, while resembling its predecessor in the matter of the vertical index spindle, possesses no other similar feature. But it has one mechanical feature which, we believe, is novel in machines of this character, namely, the absence of all springs in the mechanism for operating the index. These machines, of which several were used (and some of them are still in use), were very satisfactory in operation.

The cutting of minute pinions is now performed in

Fig. 26.—The Church Automatic Minute Pinion Cutter.

direct connection with the turning, and thereby insures axial truth. For this work an automatic machine was designed by Mr. D. H. Church, which is shown in the next view (Fig. 26). In this machine is placed a long rod of wire, which is acted upon by turning tools until a blank is formed in proper shape for cutting the teeth, the turning tools then retire and a suitable fly cutter comes into action. the blank is indexed step by step, and when all the teeth are formed, the fly cutter moves away and a cutting-off tool moves forward and severs the completed

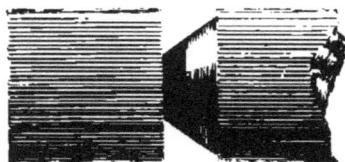

First Cut.

Cutting Off and Pointing.

Turning.

Teeth.

Fig. 27.

pinion.   The accompanying diagrams will serve to indicate some of the successive operations (Fig. 27).

The proper cutting of the teeth of escape wheels is certainly a matter of great importance and of no little difficulty; the peculiar form of the teeth demanding the utmost accuracy in workmanship. and requiring a succession of cuts by as many different shaped cutters.   It is probable that quite early in the experience of the Waltham factory it was found practicable to mount these

different cutter spindles in a single rotatable block or
head, so that the several cutters could be brought into
operative position as required.   The writer is not pos-

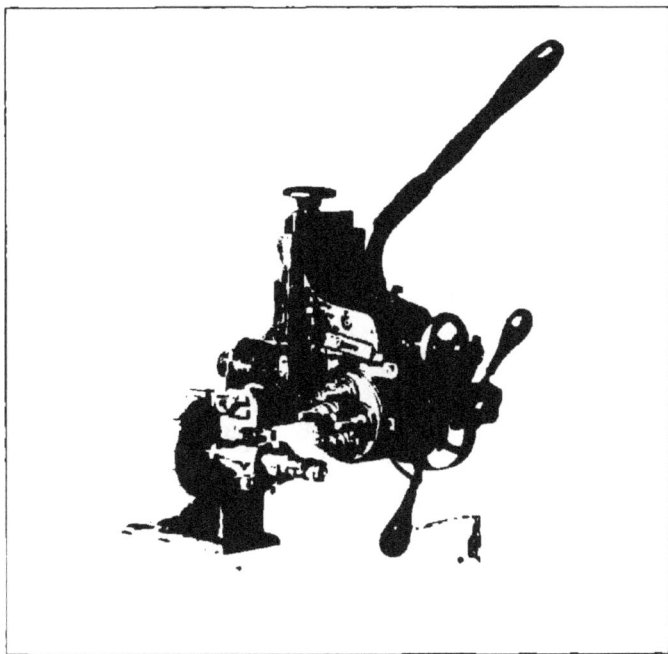

Fig. 28.—Old Escape Wheel Cutter.

itive as to the form of the machine first used for this
purpose, but it is quite likely that the machine shown
in the next view was at one time employed in that work
(Fig. 28).

The late Ambrose Webster used to take considerable pride in saying that he made the first wheel tooth-cutting machine with automatic motions which was ever used in American watchmaking. But that machine was only

Fig. 29.—Webster Escape Wheel Tooth Cutter.

claimed to be semi-automatic. It was the machine shown in the accompanying illustration (Fig. 29), and was made for cutting the teeth of escape wheels, and was automatic to the extent of moving the carriage and operating the index, and also stopping itself on the

completion of the work of each of the six cutters
required. The operator had then to bring the succeeding cutter into operative position and again start the
machine. Some years later Mr. Vander Woerd constructed a machine of a different form for performing

Fig. 30.--Woerd Escape Wheel Tooth Cutter.

the same kind of work. This machine was automatic
to about the same extent, and in the same features, as
the Webster machine, but it omitted one motion which
the Webster machine obtained, viz., the lifting of the
cutter to avoid contact with the work during its return

movement.   Mr. Woerd's machine is shown in the previous illustration.   (Fig. 30.)   To meet the requirements
of the increasing product of the factory. and at the
same time to reduce the cost of the work. the writer,

Fig. 31.—Marsh Escape Wheel Tooth Cutter.

about eleven years ago. designed another form of
escape wheel cutting machine which extended the automatic features. so as to embrace all the movements of
the machine. viz.. the reciprocating movements of the

carriage. the lifting of the cutters during the return movement. the step by step motion of the index. the successive changes of cutters. and the stopping of the machine at the conclusion of the ninety cuts required. This machine is illustrated by Fig. 31.

CHAPTER VI.

All forms of machines for the cutting of the teeth of watch wheels doubtless possess certain features in common: but while this is true it is also true that there can be. and indeed is, a great variety in forms of construction. and a diversity of ways in producing the various mechanical movements required. In the simpler or more rudimentary forms of such machines only the driving of the cutting tool would be accomplished by power. leaving the movement of the work into contact with the cutter. and the shifting of the index wheel. to be performed by the hands of the individual operator. Such a form of construction. of course. restricts the attention and labor of the operator to a single machine; and the capacity or the faithfulness of the individual largely determines the productiveness of the machine. So long as a very limited product is desired a machine of the simple form above mentioned would not only answer the desired purpose. but would. doubtless. be the most economical form to use. With increased demands. however. the cheap machine would prove to be expensive to use. It is also true that up to a certain point it is good economy to construct machines in such form as renders them capable of a variety of uses, thereby avoiding the expense of building a multiplicity of special machines which

Fig. 32.—*The Woerd Stem Wind Wheel Cutter.*

might remain idle during quite a portion of the time.

The adoption of stem-winding mechanism in watches made needful the employment of steel wheels and pinions, many of them being of other than the ordinary form of construction, and therefore complicating the matter of cutting the teeth. Experience, also, long ago demonstrated that it is not economical to attempt to remove with a single cutter all of the metal required to form finished wheel teeth. Saws are much cheaper to furnish than epicycloidal cutters, and will remove metal quite as easily. Therefore, in cutting steel wheels, a saw is used to remove a large portion of the metal, and this is followed by a forming cutter. The steel ratchet wheels, used on key-wind watches, whose shallow teeth could be formed without the removal of very much metal, were cut with a single mill or cutter, and that work could, therefore, be done with a simple machine, but when the American Watch Company commenced the manufacture of stem-winding watches, Mr. Vander Woerd constructed the form of machine shown in the accompanying illustration. (Fig 32.) This form of machine embodied two features which were of great convenience. They were, first, the incorporation of a plurality of cutters, the illustration shows three spindles, each carrying a cutter. The rear end of each of these spindles carried one member of a toothed clutch, which, when the spindle was brought into operative position, would engage the corresponding member which was carried by a running spindle or shaft. The disengaging

of the clutch and partial revolution of the spindle-carry-
ing drum were performed by means of two hand levers
(one of which is shown). A second feature which was
of very great convenience. was the provision for holding
the work at any desired angle relative to the line of
cutting, so that any of the bevel wheels and pinions
required for stem-winding watches could be cut. This
form of machine, however. was in no sense automatic in
any of its movements. but has proved to be a very con-
venient machine to have in the factory, because of its
ready adaptability.

The first automatic machine for steel wheel cutting
was designed by the writer. and. by some modifications
has been adapted to a variety of other uses. In a sim-
plified form it has been used for a number of years as a
train wheel cutter. and was alluded to in the last chapter.
In the form shown in the accompanying view (Fig. 33)
which shows a group of three. arranged to be attended
by a single operator. they are arranged to employ two
cutters. the change from one to the other being automatic. as is also the movement of the work. the turning
of the index. the lifting of the cutters during the return
movement of the carriage. and the stopping of the
machine on the completion of the work. For some
classes of work it is desired. or required. to have the

Fig. 33.—The Marsh Automatic Steel Wheel Cutters.

additional support of a tail stock, and one of these machines is shown as so provided. They, however, come short of being entirely automatic, in that they require the services of an attendant to supply the blanks and remove the completed work. Prominent in the stem-winding mechanism of the watch is the duplex or crown wheel, which wheel is in many cases provided with two sets of teeth, one of which is on the periphery and the other on the face, the latter meshing with the teeth of the winding pinion, and the former with the winding wheel. Sometimes the face teeth are located near the outer edge of the wheel so that the teeth are practically continuous with those on the periphery. It is not only possible but practicable in cutting the edge teeth of these wheels, to hold them in "stacks" on a suitable arbor, but of course in cutting the face teeth, only a single blank can be cut at a time; and great care is required to insure exact coincidence of the teeth on the face with those previously cut on the edge. The machines shown in the last two illustrations are adapted to the two classes of cutting required on crown wheels. But some years since the writer designed a machine in which both series of cutting should be performed simultaneously. The next illustration will indicate the nature and appearance of the machine (Fig. 34), which, it will

Fig. 34.—The Marsh Crown Wheel Cutter.

be seen, is in the favorite circular or continuous form,
and may be briefly described as provided with four
cutting spindles, each of which is adjustable in two
directions. The first two of these are for cutting the face
teeth, and the other two for the edge teeth. The first
of each set carrying a saw, and the other two the forming or finishing cutters. These four spindles are
mounted on double slides, and are suitably disposed
around a central carriage or turret, which turret carries
five vertical quills, in which are the work-carrying
spindles: the wheel blanks to be cut being held by suitable chucks at the top, while the lower end of each
spindle is provided with an index, which is accompanied
with a holding latch and a suitable arm and pawl for the
step by step movement of the index. It will be observed
that while there are but *four cutter spindles* there are
five work holding spindles. This fact allows the operator
to place a blank in the chuck of the fifth spindle while
the work of cutting is progressing on the other four, so
that the work of the machine is continuous. After the
blank has been secured in place, the turret turns one-
fifth of a revolution, which carries the blank into position to be acted upon by the first cutter, which is in the
form of a saw, and as the index is operated step by step,
the saw passes down and out at the proper angle to the

face of the wheel, each movement removing a portion
of metal, thereby blocking out the face teeth. In the
meantime the operator places a blank on the second
spindle. When all the face teeth in the first blank have
been blocked out, the turret again makes a partial revolution, which carries the first blank to position to be
acted upon by the finishing cutter, and the second blank
to the position just vacated by the first. The third position is for the saw for the edge teeth, and the fourth is
for the finish of the edge teeth. The next movement of
the turret brings the finished wheel to its starting place,
when it is removed and another blank put in its place.
All the operations proceed simultaneously, so that a
completed wheel which requires the four milling cuts, is
produced during the time occupied by one operation,
and inasmuch as the somewhat large number of teeth in
the wheel necessitates a corresponding length of time for
cutting, the operator is able to attend another machine
also. The latest form of wheel cutter, which is shown
in the next view (Fig. 35), is the invention of Mr. D. H.
Church, and, by a suitable arrangement or combination
of cams, can be adapted to the cutting of stacks of
wheels (as shown in the cut), or to the cutting of single
wheels on the face. This machine, by reason of its
automatic feeding mechanism, requires no individual

operator, but allows him or her to attend to a large group of similar machines, which may be employed on

Fig. 35. - The Church Automatic Wheel Cutter.

a great variety of cutting. The use of a liberal amount of oil allows these machines to produce an unsurpassed amount of work.

CHAPTER VII.

It will be understood that in this work it is not intended to describe. nor even mention. all of the various machines used in the manufacture of watches, but rather to select a few types, and note the progress in their productiveness. which the growth of the business has demanded. and which has been a large factor in the reduction in manufacturing cost. and of which the great public has reaped the larger benefit.

Next to the plates. and the wheels and pinions. the numerous screws required may be regarded as prominent. We will. therefore. devote a short chapter to the consideration of some of the various forms of screw-making machines which have been in use in the American Waltham Watch Factory. The original threads. or rather. the threads used in the early Waltham watches are said to have been obtained from Swiss "jam plates." and when. in later years. definite pitches for all the sizes of screws were determined upon. they were established on the inch measurement. which system was in use in the factory previous to the adoption of the "metric system" which is now used. So that while the various

numbers of threads per inch were somewhat systematic
(varying from 110 to 240 per inch) yet their equivalents
in metric measurements *seem* not to be so systematic.

The early method of screw-making, consisted of
the use of a small bench lathe, with the ordinary split
chuck for holding the wire rods, which chucks were
closed by means of the regular draw-in-spindle and hand
wheel. On the lathe bed was a double slide rest with
one tool for turning down the wire to form the body of
the screw, and another tool for "cutting off." The
lathe was also provided with a swing, or "tumble tail
stock" containing two or more spindles, one of which
served as a "stop" to regulate the length of the screw,
another spindle carried the threading die. For turning
the lathe during the threading process, the hand of the
operator was employed, so that care was required in
running on the die to avoid twisting off the slender
screw when the die came in contact with the shoulder
which formed the under side of the head. After the
threading operation the cutting off tool was brought into
action, and the wire rod partially severed, enough metal
being left to sustain the screw. At this point the operator would pick up a "slotting plate," and holding it to
the nearly severed screw, turn the lathe and run the
screw into a tapped hole in the plate until it came in

contact with the head. when the severing process was completed by twisting it off. This method of screw-making in the Waltham factory was so long ago discarded. as to make it a matter of some difficulty to

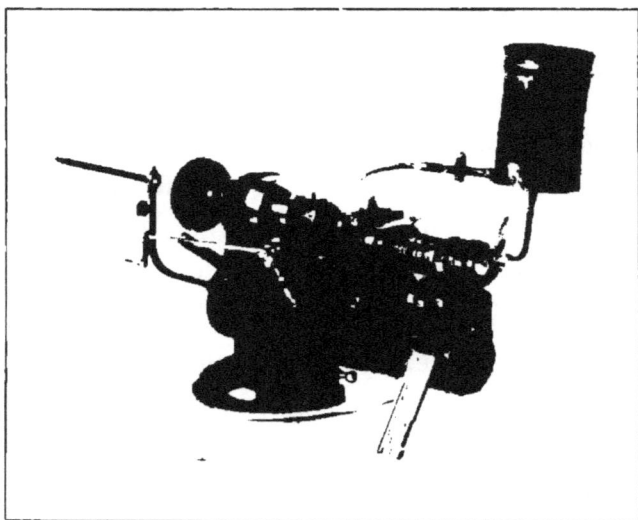

Fig. 36. *Early Form of Screw Cutting Lathe.*

gather the material required for an illustration, but the accompanying view will serve to show what was at one time in use. (Fig. 36). Mention has been made of the "slotting plate;" in this view such a plate is shown leaning against the bed of the lathe. When all of the two rows of holes in the slotting plate had been filled with

screws in the manner above mentioned, it was fastened
to the carriage of the slotting machine, which, on being
started, would steadily carry one row of screw heads
into contact with the running saw, the workman in the
meantime taking another plate and continuing his work
of turning, threading and cutting off.   When the first
row of screws were slotted, the carriage was drawn

Fig. 37.—*Early Form of Screw Head Slotting Machine.*

back to its former position, the plate removed and
reversed, and the other row slotted as before.   The
plates were then taken by a boy who removed the
slotted screws, and returned the plate to the workman
for another filling.   The next figure shows one of
the above mentioned slotting machines (Fig. 37).
The brass screws to be used in the rims of balances,
while made in a similar, but not identical manner as that

just described, were taken off in a sort of block instead of the above mentioned slotting plate. The reason for this method being the readiness of releasing the slotted screws from the slotting block, or holder, which was desirable from the fact that the threads of balance screws are very fine and delicate, and therefore liable to

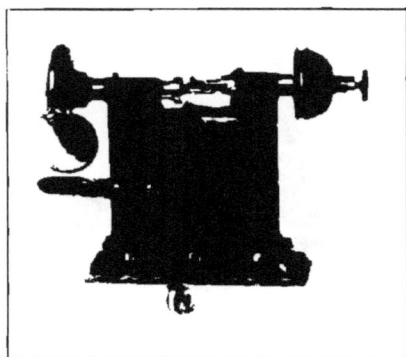

Fig. 38—Old Balance Screw Slotting Machine.

injury. This slotting block was circular in form, and consisted of two plates held together by a screw nut. The joint between the two plates was drilled with a row of radial holes, tapped to fit the balance screws, which were inserted, as made, in a manner similar to that already described When this block was filled it was placed on the arbor or spindle of the special slotting machine shown in the next view. (Fig. 38), which also shows the slotting block. On starting the machine the block was slowly revolved. carrying the heads of the radiating screws into contact with the slotting saw. When completed the block was removed from the machine. the clamping nut turned back to allow the two halves of the block to separate. when the screws would readily drop out.

The large number of screws required in watchmaking, when carried on to the extent which it had been for years at the Waltham factory. led to the attempt to improve upon the above described methods. and to Mr. C. V. Woerd belongs the credit of taking the first step in that direction. The next view (Fig. 39) shows the first Automatic Screw-making Machine. This was made in 1871. and has been in constant use ever since that time. This particular machine was designed for making "Jewel Screws." this screw doubtless being selected for two reasons; first. because of the large number used, and second, because being so very small. it would be more easily made by a machine. About four years of successful use of this machine led to the designing of a larger machine. substantially the same in principle, but adapted to the production of the larger watch screws. This heavier machine is shown in Fig. 40. Examination of this picture will show that a

Fig. 30.—Ward's First Automatic Screw-Making Machine.

Fig. 10 Second Automatic Sugar-Making Machine, for Larger Sizes

machine of this kind must of necessity be expensive to build; and this fact led to the designing of a machine more compact in form as well as more simple, and therefore of cheaper construction. This latter form, which was designed by Mr. D. H. Church, possesses some features, besides cheapness, which are not found in the Woerd machine, and which enable it to perform some kinds of work which it would be difficult, if not impossible, to do on the older machines. The Church machine is shown in the next view (Fig. 41.)

In all the foregoing machines the various and requisite operations of loosening the chuck, feeding forward the wire rod, tightening the chuck, turning the body of the screw, cutting the thread, and severing the blank from the rod, are performed *successively*, while the slotting of the head could proceed during the time occupied by one or more of the other operations.

It was the conviction of the writer that a great economy would be secured by the adoption of a machine so designed that *all* of the above operations could proceed simultaneously. Such a machine was made, and has for ten years been employed in making the largest screws used in full plate watches, making pillar screws at the rate of twelve per minute. In a slightly modified form these machines are used for making balance screws,

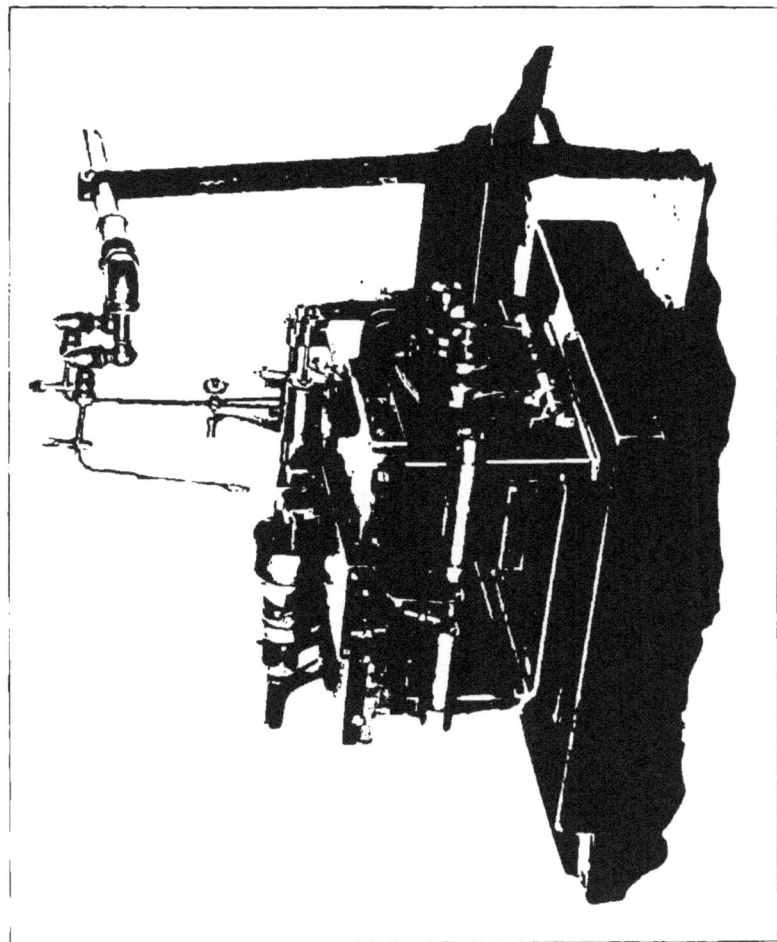

Fig. 41. The Church Automatic Screw-Making Machine.

which, being of brass, are capable of more rapid production, so that twenty screws per minute are made on a single machine. (Fig. 42).

But a single operative can readily attend to six or more of any of the above mentioned forms of automatic machines, so that while by the older method a man might make 1,200 to 1,500 screws per day with a little aid from a boy, it is possible for one man alone with these machines to easily make 50,000 to 60,000 per day. This result is attained by the adoption of the automatic features, by which it is made possible for one person to attend to a number of machines, and also the arrangement of the machine which provides that all the successive operations involved in making the screws are carried on simultaneously. This latter fact is accomplished by the use of a multiplicity of spindles, each of which carries a rod of wire, and which successively present themselves to the various tools, in the favorite "procession," which continues until the rods of wire are exhausted, when the machine will stop until replenished.

But the turning, threading, cutting off and slotting of the screw does not complete it, for there remain the finishing operations, which involve much more of expense than attends the making operations. Formerly, it was

Fig. 42. - The Marsh Automatic Soccer Making Machine

needful to handle each individual screw as many times as there were distinct operations involved in the finishing process. But improved processes and machines have reduced the expense of finishing the heads to quite an extent.

The next view shows a lathe once used for finishing screw heads, (Fig. 43). The running spindle carried a chuck, whose center was threaded to receive the screw, which was held in the fingers of the operative, and when properly presented to the revolving chuck, it would be drawn in, till the screw head came in contact with the end of the chuck. The operative then applied the surface of a fine oil stone to the head of the revolving screw, and moved it back and forth until the surface of the screw head was carefully ground; following this came a stick of boxwood whose surface was charged with Vienna lime and alcohol, a similar manipulation of this stick would produce the desired glossy surface. The operator then applied a suitable screw-driver to the slot of the screw head, and, by the left hand, turned

Fig. 43.—Old Screw Head Finishing Lathe.

the lathe spindle backward. unscrewing the finished
screw from the chuck.   To insure the proper position
of the above mentioned oil stone and stick. when in
operation. a suitable rest was provided.

In the case of the brass balance screws. which must of
necessity be entirely uniform. to insure poising, a double

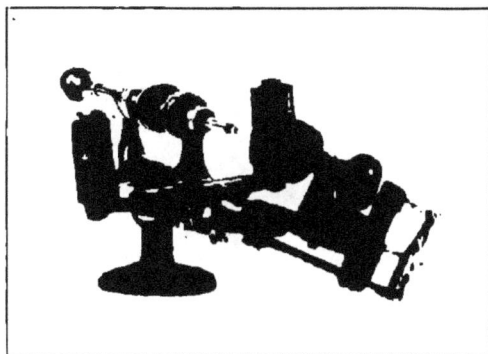

Fig. 44. Balance Screw Head Finishing Lathe.

rest was provided, and a file was also employed to
bring the heads to the requisite length.   In this machine
(Fig. 44) an attachment was used for polishing the sides
of the head as well as the end.   The later practice is to
grind and polish the ends in large numbers at a time. and
for finishing the sides of the head. the machine shown in
the next view was designed (Fig. 45).   This will be
recognized as one of the continuous running type of

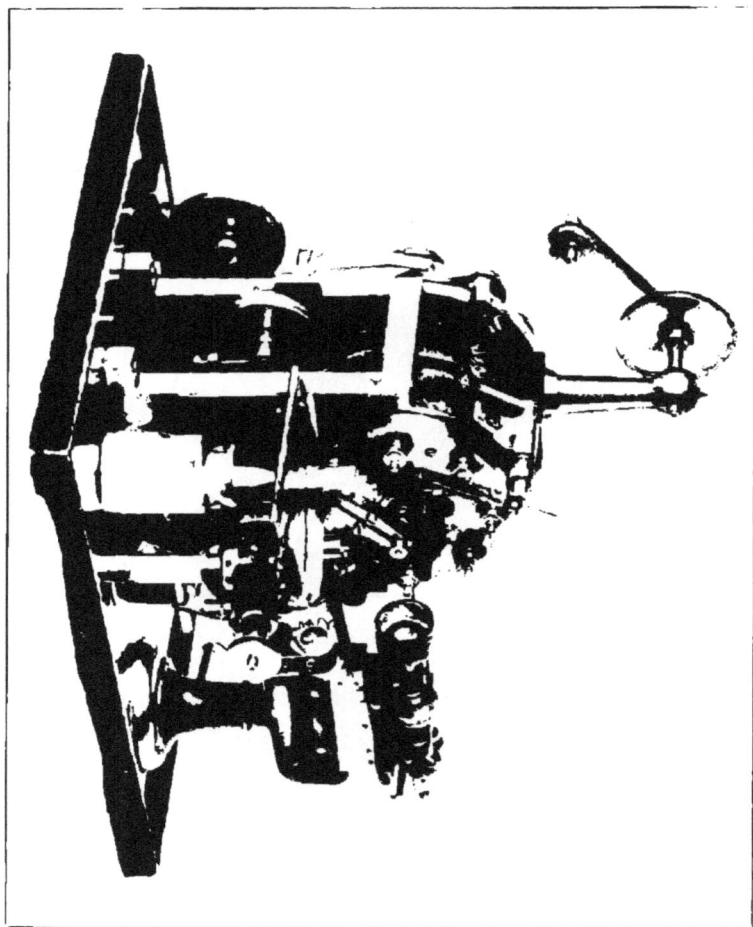

Fig. 45.—Marsh Continuous Screw Head Finishing Machine.

machines previously mentioned. This machine has a revolving head, containing eight spindles, each of which is provided with an interior threaded chuck. The head revolves by a step by step movement, and the spindles, when they successively reach certain positions, are put in rapid motion.

When in one of these positions the operator presents the point of a screw to the chuck, and it is at once screwed in. That spindle then moves to another position, and in its progress passes under the action of a fine file, which removes any burr which may have been made by the slotting saw. At a later period the screw reaches a position to be acted upon by a swiftly revolving wood lap, which also has a vibrating motion. Another step or two carries the screw head to the action of a buff wheel, which gives additional gloss to the surface of the metal. At the next step the screw disappears, so that when the spindle reaches its first position the chuck is empty and ready for another screw; and so on. Another form of screw head finishing machine was originally made for finishing the minute screws used in holding jewel settings. Some girls acquire such skill in handling these screws that they could readily put them into the chucks of the running spindles, but it was entirely a matter of feeling.

A later form of machine has made this skill unnecessary, as each spindle, when it reaches the proper position, receives its screws automatically, or if it fails to

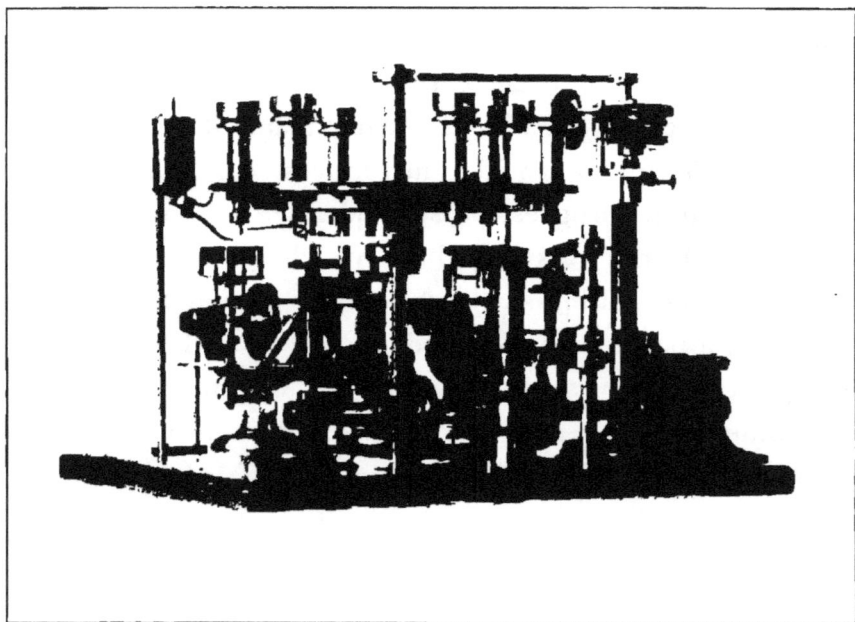

Fig. 46.—The Wood Machine for Finishing the Heads of Jewel Setting Screws.

do so, that failure stops the machine, and so avoids
injury to the empty chuck by the action of the grinding
and polishing mechanism. This machine (Fig. 46) is
the invention of Mr. Gleason Wood. Other forms of

machines have been made for finishing round-topped heads, but this chapter is quite long enough, and we will not attempt any description of them.

CHAPTER VIII.

In the last chapter we very briefly described the machines for the making and finishing of the balance rim screws.

In this chapter we will review some of the machines which have been. or are now employed in the production of the balance itself. The making of the balance involves a larger number of successive operations than any other single piece of the watch movement. It also demands the utmost care in manufacture. to insure its absolute truth and reliability of action under the varying conditions to which it is subject when in the performance of its important and delicate work.

It is to be understood that we refer to the bi-metallic or expansion balance, (which is the only form of balance used by the American Watch Company) and that from the nature of its structure it is absolutely essential that it be as near perfect as possible in every particular. We are however to consider. not the balance itself. but the machines used in its production; we will begin with the operations required after the brass portion has been united to the steel body of the balance blank.

For the satisfactory action of the completed balance under varying temperatures, it is desired that the more expansive brass be made as dense as possible. This density is ordinarily secured by some sort of a compression process. Quite possibly this may at some time

Fig. 47 —Early Machine for Hammering Balance Brass.

have been accomplished by a long succession of blows by a hand hammer. if so, the operation must have been a tedious one. and probably not entirely satisfactory in its results.

The first mechanical means known to the writer for performing this work was in the form of a small trip hammer designed for this special use (Fig. 47.)

This machine was provided with two grooved rolls mounted on and near one end of two parallel shafts. which near their other extremities carried rolls of differ-

Fig. 48.—Another Early Machine for Compressing Balance Brass.

ent form.   By means of two little belts one of these shafts was made to drive the other, the first one receiving its motion through a belt from a countershaft.

A count wheel was so arranged as to receive a step by step movement by means of an arm and pawl attached to the axis of the trip hammer.

The balance blank to be operated upon was held in position by a detached arbor which at one end was made slightly tapering and of such size as to fit the center hole of the balance. The other extremity of this arbor carried a role or disc of substantially the same diameter as the balance blank.

When in position the blank rested on the periphery of the two grooved rolls. and the disc rested in a similar manner on the two belted rolls at the other end of the parallel shafts: a suitable spring was arranged to maintain sufficient pressure to compel the motion of the two drive rolls to rotate the balance arbor. The balance blank being in position. the machine was put in motion and the feed rolls would slowly revolve the blank.

At every stroke of the hammer the pawl would move the count wheel a single tooth. and by means of a suitable stud attached to the wheel it would at a proper time lift a latch and release a shipper lever and stop the machine.

The object of this counting mechanism was to insure uniformity in the hammering of the rim, so that the

Fig. 49.—Self-Feeding Automatic Compressor for Balance Brass.

metal on the entire periphery of the balance blank
should be of uniform density.

This form of machine was succeeded by the one
shown in (Fig. 48) which, while more efficient in every
way than the hammering machine, was displaced by
the self-feeding and automatic acting machine shown
in (Fig. 49.)

In this machine the blanks to be compressed are
placed in the feed tube whence they are taken one by
one from the bottom of the pile and carried forward
and deposited on a suitable lifter which elevates the
blank to a position where it can be grasped by the
grooved rolls which gradually but relentlessly close in
upon it. not "to squeeze the life out of it," but to give
it a greater life. When the blank has been compressed
to a definite size the rolls retreat and allow the compressed blank to drop, when it is pushed out of the way
by a suitable arm, just in season to make way for
another victim, which is "put through the mill" in the
same manner.

This machine has only to be kept loaded with blanks
and itself "does the rest."

The facing and recessing of the blanks then follow;
and these operations were formerly done in ordinary
bench lathes, each of which required an operative

(Fig. 50.) The later practice is to employ an automatic lathe, which takes the blanks from a loaded tube and automatically places them in the chuck, when the cutting tool advances and does its work and then retires.

Fig. 50.—Early Balance Facing and Recessing Machine.

when the chuck opens, the turned blank is ejected, and a new one received, and so on. The attendant, who can care for a number of machines, has for his principal work the sharpening and renewing of the cutting tools (Fig. 51.)

Fig. 51.—The Church Automatic Facing and Recessing Machine.

The blanks being turned and recessed, there follows the removal of two sections of the thin web of steel, leaving a transverse bar of metal, forming the arms of the wheel.   As this bar is widest at the middle, tapering

Fig. 52.—*Balance Webb Cutter.*

on both sides towards the rim, the removal of the two sections of web is not so simple as it would otherwise be.   Years ago it was the practice to make four cuts through the web with a suitable shaped mill which was sunk into one side.

The next operation was performed in a lathe, the spindle of which was given a reciprocating motion, and in connection with it a special crossing tool was made to cut through the web near the rim, thus completing the rim. (Fig. 52.)

The modern practice is to remove both sections of the web by means of a specially constructed punch and die.

The drilling and tapping of the numerous holes in the balance rim, for the reception of the adjustment screws, is unavoidably an operation of considerable extent, inasmuch as from fourteen to eighteen holes in each balance must be accurately located and carefully drilled and tapped at the proper stage of the work, and retapped when the balance is finished.

The earliest form of machine known to the writer for this drilling, consisted of a bench lathe with a swing tail stock containing a number of spindles, each carrying a special form of tool. In the running spindle of the head stock was a chuck, in which the balance was mounted, and provided with a transverse spindle, the outer end of which carried an index, graduated to correspond to the desired number and position of the holes in the balance.

The inner end of this little spindle carried a face plate to which the balance was clamped, and in such position as to bring the desired location for a hole in the axial line of the lathe spindle. By having the chuck carefully counterbalanced it could be safely revolved at sufficiently high speed to drill the holes, etc. This lathe is shown in (Fig. 53.)

The transition from the slow and somewhat expensive

Fig 53.—Early Balance Rim Drilling Machine.

method of drilling just described. to a method both rapid and correspondingly cheap, was in this case a radical one. omitting the steps of gradual progress which has marked improvements in most directions. Yet there was at least one preliminary step in a direction allied to these operations. The operation of tapping the numerous holes was

one of some delicacy and yet demanding rapidity in order
to avoid undue expense; the speed was secured by mount-
ing the tap in a small running spindle driven by power,
but great watchfulness was demanded to avoid running
the threading tap in too far, before reversing the direction of revolution.

The form of tapping machine shown in the next view

Fig. 54.—*Early Balance Tapping Machine.*

(Fig. 54), was therefore designed, which insured both
speed and safety, inasmuch as it was arranged to give the
tap a yielding pressure when entering the hole, and also
allowed only a definite number of forward revolutions.

This tapping device was incorporated by Mr. C. V.
Woerd into an automatic drilling machine, in the form
shown in the next view (Fig. 55), and being automatic,

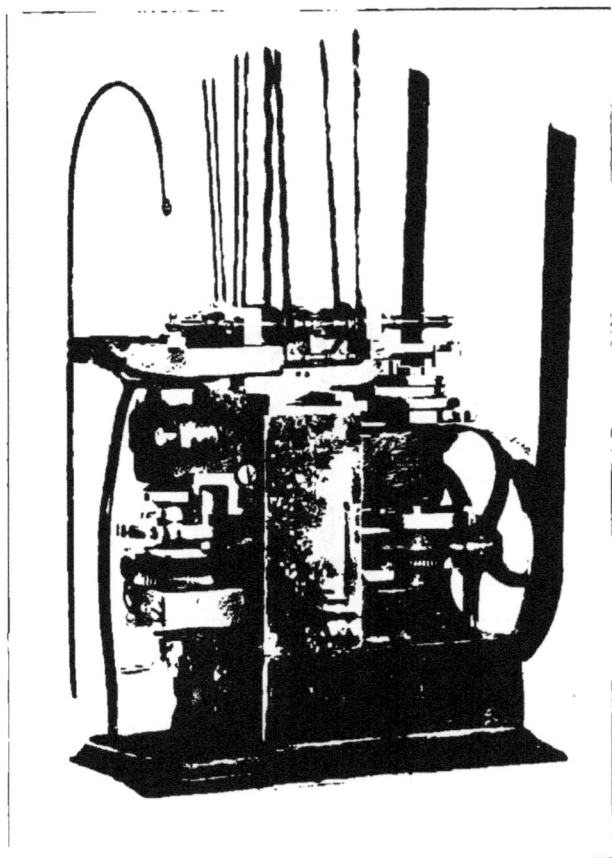

Fig. 55.—Woerd's Automatic Balance Drilling Machine.

save only in the supplying and removing of the balance. two persons are able to attend to twelve machines or more. which are arranged in close order.

Fig. 56.—Marsh Modern Balance Tapping Machine

For the final tapping of the holes, the machine shown in the next view has proved to be most excellent, as the entire number of holes are tapped simultaneously (Fig.

56). The finish turning and glossing of the balance rim is an operation demanding special care and accuracy, absolute truth and accuracy in diameter being required. The earlier practice was to bed the balance in cement on

Fig. 57.—Finish Turning and Glossing Machine.

chuck, which was provided with true running arbor or pin. As the cement required to be softened by heat, both when securing the balance to the chuck, and when removing it after being turned, considerable time was necessarily consumed in securing the balance and getting it in readiness for the turning operations. At one time

the turning tools were in removable holders, which were
taken in succession by the workman, and laid on a hard-
ened tool-rest, in front of the work, and carefully passed
over the running balance, each tool removing a small

Fig. 58.—Marsh Modern Finish Turning and Glossing Machine.

portion of the metal.   A later form of machine, or rather
a modification of the foregoing machine, is shown in the
next view (Fig. 57).   This shows the turning tools
mounted in a holder which is arranged to swing on an
arbor, and also to slide on the same arbor, the holder

resting on a suitable guide; the movement of the slide being imparted through a suitable hand lever.

The illustration also shows the revolving polishing disc. for giving the finishing gloss to the rim.

This form of machine has been superceded by the one shown in the next view (Fig. 58), which is another application of the " procession " idea.

This machine is arranged with a turret carrying four running spindles. each of which at its upper end carries a specially made chuck for holding the balances to be turned. Suitably disposed around this turret are three tool rests, two of which are provided with turning tools. while the third carries a revolving polishing disc. This arrangement allows of the simultaneous performance of the two turning operations. and also of the polishing, while the fourth spindle is receiving a fresh balance, which will duly follow the others step by step around the circle, each spindle when arriving at the starting position bringing with it a completed balance; which the attendant will remove and supply its place with a fresh one. This arrangement enables one man to accomplish at least two and one-half times as much as by the previous method.

E. A. MARSH.

CHAPTER IX.

In the last chapter we reviewed some of the special machines used in the manufacture of balances; and it may not be inappropriate at this time to consider the companion of the balance—the hairspring, and its stud and pins.

It is probable that hairsprings are made in the same general way, by all makers, while the tools and machines employed in any one factory may be, and doubtless are, entirely dissimilar in form to those used for a like purpose in other factories. Of these various tools we can make mention of but few, and of them it will not be possible to illustrate the successive forms or steps in the long process of evolution.

We believe that there was a period in the history of this factory when hairsprings were purchased, instead of being manufactured. At a later period the finished wire was purchased, from which the springs were made. But about twenty-five years ago a machine for forming the wire was built. Subsequent modifications and improvements have resulted in the production of the machines shown in Fig. 59. These machines are so far

Fig. 50.—Modern Automatic Hair Spring Wire Forming Machine.

automatic in operation that one man only is required to keep them busy. The character of the work done by the several machines is modified by the successive requirements of the wire, and include the drawing of the round wire to the exact diameter desired, then flattening of the wire by repeated rollings. and lastly by repeated and careful drawings through the finishing dies, by which the exact dimensions desired are obtained. together with the smooth and glossy surface which is indispensable to the production of a highly finished and bright colored spring.

The coiling of hairsprings seems to belong among the class of mechanical operations, or manipulations, which are not susceptible of marked improvement. There may be obtained a measure of superiority in the quality of the tools employed, but the processes of production admit of little variation. One exception may be made to the foregoing statement, it being the method of forming the overcoil of Breguet springs. which method was devised by the late John Logan. and by which it is possible to so form and confine the over-coil that it can be tempered complete, not requiring the careful and somewhat tedious manipulation otherwise demanded. The round hairspring stud is one of the small parts of a watch movement, and quite simple in form as compared with the

older pattern, which was a long wedge like a piece of
steel which was attached to the watch plate by a screw
and two steady pins, and projected toward the spring
like a balance cock.   But, although quite simple in form,
the diminutive size of the round stud makes it an incon-

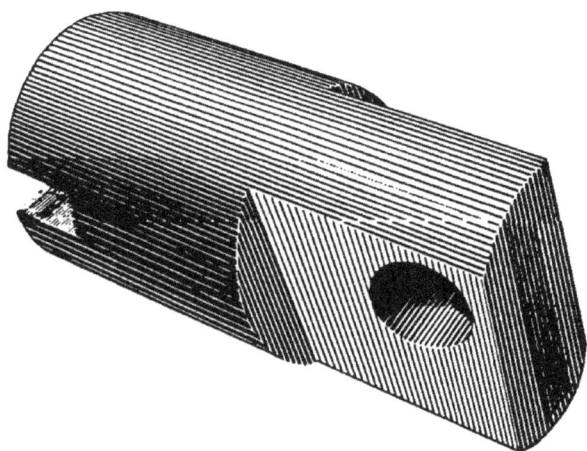

*Fig. 60.—Form and Relative Dimensions of Hair Spring Stud.   About Twenty-
five Times Actual Size.*

venient piece to manufacture—at least this was true
during the early years of its production.   The accompanying diagram (Fig. 60) will serve to indicate the
form and relative dimensions of the studs, the largest of
which are less than 4-100 of an inch in diameter, by
about 9-100 of an inch in length, and at one end is flat-
tened on two sides so as to form a sort of tongue, near

one corner of which is drilled a hole to receive the outer end of the hairspring.   The opposite end of the stud is provided with a slit to receive a screwdriver.   At one time it was the practice to cut off from a rod of wire, pieces of suitable length for one stud each.   Then followed the operation of milling one of the ends, and finishing the other to the proper length.   Next came the drilling of the hole for the spring.   It is essential that this hole should be at an exact right angle to the axis of the stud, but with so little of total length, as compared to its diameter, it proved a very difficult piece to hold during the drilling operation, so that when inspected a large number were discarded for imperfections.   But all that is a thing of the past.   A special form of continuous running automatic machine, shown in Fig. 61, now does the work of milling the two sides, drilling and broaching the hole, and cutting to exact length, and producing a stud every six seconds.   This is another instance of the efficiency of the "procession," or continuous running system.

The writer was once told by Mr. Chas. W. E
Patentee of the Safety Pinion, that when in
was an apprentice to learn the
one occasion the boss was
he therefore told the you

employ himself making hairspring pins. "Well, how many shall I make?" "O, make about a pint." Hairspring pins are not bulky articles, and filing them by

Fig. 61.—Marsh Automatic Hair Spring Stud Machine.

vise, twirled by thumb and that was the *old* way. An the wire was held in a

spring chuck which was in a spindle removably connected with a spindle driven by a belt.   With this was
also employed a rotary file, as a substitute for, and
improvement upon, the common hand file.   This method,
while probably producing better and more uniform pins
than the primitive method, was by no means rapid, inasmuch as it was required to remove the inner-spindle
from the outer driving spindle, and after cutting off the
completed pin, to loosen the chuck and pull out the wire
to a distance required for another pin.   Only a few
inches in length of wire could be handled in this
machine.   But for several years past automatic machines
of the form shown in Fig. 62 have been used.   These
take pieces of wire about 30 feet long, and produce pins
at the rate of 20 to 35 per minute, according to the
material used.

The proper adaptation of the hairsprings to the balances of watches is an indispensable requisite in the
obtaining of a correct time rate.   It is a matter of absolute exactness, and therefore one demanding special care
and accurate tests.   Until within comparatively a few
years this work was done by the " cut and try " method.
that is. by repeated trials of different springs, until one
was found which would meet the requirements of each
individual balance.   The testing or trial could not be an

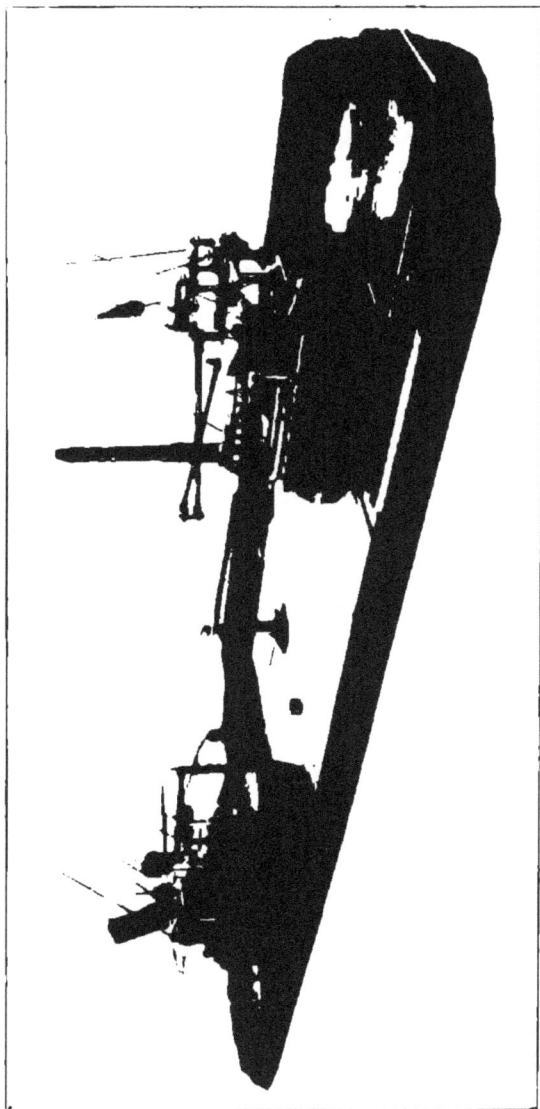

Fig. 62.—Marsh Automatic Hat Spring Stud Pin Machine.

instantaneous matter, and often involved many changes;
but in the case of the ordinary flat hair springs admitted
of such latitude as could be corrected by "taking up"
or "letting out" the spring.

Extended and careful tests have demonstrated that the
correct action of a Breguet hair spring demands that the

Fig, 63.—*Device for Testing Balances and Springs.*

stud be applied at a certain point in the overcoil, and
that only a very slight deviation from this position can be
allowed without destroying the proper action of the
spring.

The late John Logan, who was extensively known as
a maker of hair springs, was for several years employed
as a " watch springer," and gave a great deal of thought
to the problem of cheapening the cost of his work.   As

a result of his thought and study, he invented a system
of testing all hair springs by a standard balance, and all
balances by a standard spring, and grading the springs
according to their relative strength, and, by means of a
long studied and carefully prepared schedule, or table,

Fig. 64.—*Logan Device for Testing Balances and Springs.*

selecting the springs adapted to the various balances.
This scheme was the occasion of a long and expensive
lawsuit, resulting in favor of Mr. Logan. Some of the
various devices for testing balances and springs are
shown in the accompanying illustrations, Figs. 63 and 64.

Mr. E. J. Hall devised and patented the little compara-
tors shown in Figs. 65 and 66, which are of value to a

certain extent. but their use demands an amount of keen
and accurate perception which not every workman
possesses.

Fig. 67 shows a special form of balances. or weighing

Fig. 65.—The Hall Patent Comparator.

scales. by which the "avoirdupois" of the complete
watch balance could be determined.   This was devised
by Mr. Thomas Gill, who also designed the very ingenious and useful gauge shown in Fig. 68 for weighing the

relative strength of springs. By the use of these two devices, an *approximate* selection of springs, as to their adaption to balances could be made. This system, while helpful to a limited extent, was far from perfect, as it was manifest that while the *total mass* of each of a large number of balances might be exactly equal in weight to that

Fig. 66.—Another Hall Patent Comparator.

of each of the others, yet the metal might be so disposed as to give vibrational weight in some cases much greater than in others. So that the only accurate means for ascertaining the effective weight of balances is by vibration. The same method is also required to measure the strength of hair springs with the accuracy required. Such careful and minute gauging of necessity demands

that the determining vibrations shall be continued through
an interval sufficiently prolonged to disclose or reveal the
peculiar condition of each individual balance or spring.

To do this work thoroughly without entailing a large

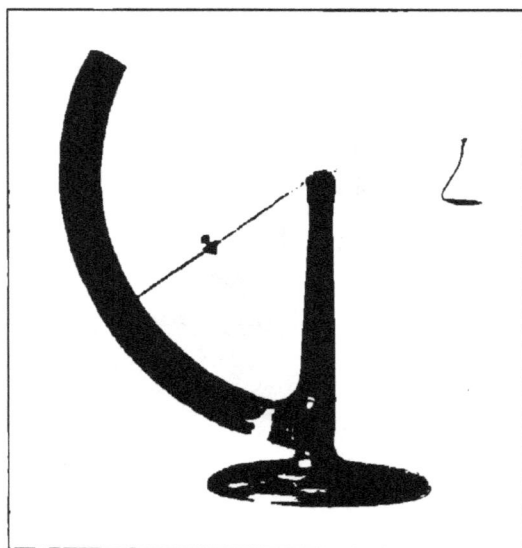

Fig. 67.—*The Gill Scale for Weighing Balances.*

additional expense, a form of special machine was
devised, which has proved to be very efficient and relia-
ble. Fig. 69 shows a number of these machines as
arranged for use, some of them fitted for vibrating balances, others for springs, and still others for testing. By

means of this system the springs and balances are
selected entirely independent of the movements to which
they have been assigned, so that they do not come
together until they reach the hands of the finisher.

In closing this series of papers, which have extended
far beyond the original intention of the writer, it is

Fig. 68.—The Gill Gauge for Determining the Relative Strength of Hair Springs.

proper to say once more that it is impossible to convey
more than a general idea of the direction and nature of
improvement in machinery for watchmaking, and to
indicate some of the steps in which progress has been
made. The list of improved machines is by no means
complete. No mention whatever has been made of

some of the most recent changes. At no time in the history of this company have changes been so numerous nor so radical as within the past two years. Enough, however, has been written to show that very great changes in manufacturing methods have been in progress during almost the entire existence of this factory. Many of the earlier methods and machines now seem crude; possibly they were known to be such at that time, but it is difficult for us at this day and in the light of recent mechanical achievements to realize the primitive conditions of forty years ago.

Perhaps no one of the past two score years has failed to witness some degree of improvement in machinery. Doubtless succeeding years will also bring additional changes, but it is not probable that the future student of the history of watchmaking will be able to discern in any like period as much of progress as will mark the last decade of the present century.

We have aimed to write of methods and machines, and not of men. But it seems proper in these closing words to make mention of two who are deservedly prominent; the first as being to a certain extent a pioneer in the field of designing and building watchmaking machinery, and the second who has by his fertility and originality in the field of invention, achieved so much in the embodiment

of automatic features as render his recent machines won-
ders of mechanism.   We refer to Mr. C. S. Moseley and
Mr. D. H. Church.